Handbook of Starch Hydrolysis Products and their Derivatives

Handbook of Starch Hydrolysis Products and their Derivatives

Edited by

M.W. KEARSLEY
Business Development Manager
British Sugar Technical Centre
Norwich

and

S.Z. DZIEDZIC
Business Development Manager
British Sugar Technical Centre
Norwich

BLACKIE ACADEMIC & PROFESSIONAL
An Imprint of Chapman & Hall
London · Glasgow · Weinheim · New York · Tokyo · Melbourne · Madras

Published by
Blackie Academic & Professional, an imprint of Chapman & Hall,
Wester Cleddens Road, Bishopbriggs, Glasgow G64 2NZ

Chapman & Hall, 2–6 Boundary Row, London SE1 8HN, UK

Blackie Academic & Professional, Wester Cleddens Road, Bishopbriggs, Glasgow G64 2NZ, UK

Chapman & Hall GmbH, Pappelallee 3, 69469 Weinheim, Germany

Chapman & Hall USA, Fourth Floor, 115 Fifth Avenue, New York, NY 10003, USA

Chapman & Hall Japan, ITP-Japan, Kyowa Building, 3F, 2-2-1 Hirakawacho, Chiyoda-ku, Tokyo 102, Japan

DA Book (Aust.) Pty Ltd, 648 Whitehorse Road, Mitcham 3132, Victoria, Australia

Chapman & Hall India, R. Seshadri, 32 Second Main Road, CIT East, Madras 600 035, India

---

First edition 1995

Typeset in 10/12pt Times by Cambrian Typesetters, Frimley, Surrey

Printed in Great Britain by The University Press, Cambridge

ISBN 0 7514 0269 9

A catalogue record for this book is available from the British Library
Library of Congress Catalog Card Number: 95–79038

∞ Printed on permanent acid-free text paper, manufactured in accordance with ANSI/NISO Z39.48-1992 (Permanence of Paper).

Preface

There is an ever increasing demand for sweeteners of all types for the food and drink industry. These include bulk sweeteners derived from natural sources, such as glucose syrups from starch and sugar from sugar beet or cane, as well as the intense synthetic sweeteners. Whilst both sugar and artificial sweeteners are subject to periodic vilification in the popular press, starch-based products do not attract the same poor publicity and this is undoubtedly one of the major reasons for their increased use in foods and drinks. There is thus a continuing need to ensure that up-to-date technological and physiological information on glucose syrups and their derivatives is available for reference by those involved in food research and development.

The contributors to the individual chapters in the book are all recognised experts in their fields and provide a unique insight into their individual speciality. This book describes the different starch sources used commercially to produce glucose syrups leading onto the manufacture of these products, principally by enzymic means. Further chapters describe the physico-chemical and physiological characteristics of glucose syrups and related products, such as maltodextrins and the starch derived polyols, and the use of these products in the fermentation industry. For those wishing to add a more quantifiable aspect to their work, the chapter on analysis of glucose syrups will provide such information.

A chapter on the applications of glucose syrups completes the book.

The information in this book will be invaluable to all such scientists and technologists to enable them to make the optimum use of these products in the course of their new product formulations. It will also be essential to the food engineer who may require to know viscosity data or other physical properties. For the research oriented, the book provides a comprehensive list of reference material to facilitate further reading should the reader wish to obtain a more detailed insight into these versatile products.

M.W.K.
S.Z.D.

Foreword

Glucose syrups are arguably the most versatile of all food sugar ingredients. They have an established reputation covering more than forty years of regular commercial use and have been reviewed on previous occasions (Birch *et al.*, 1970; Dziedzic and Kearsley, 1984). The traditional industrial index of characterisation is the 'dextrose equivalent' (DE) which has now been largely replaced by more detailed specifications of composition based particularly on chromatographic analysis. Glucose syrups can therefore be designed to meet many different nutritional and technological requirements.

A principal use of commercial glucose syrups has been as a carbohydrate sweetening substance. Paradoxically however, this originally placed the glucose syrups at a disadvantage compared to sucrose which is 3–4 times sweeter than regular starch hydrolysates. The ability to isomerise high DE syrups to 'high fructose syrups' has now redressed the balance and high fructose glucose syrups have become a major economic ingredient in the soft drinks industry.

A further disadvantage of starch hydrolysates compared to sucrose is their involvement in non-enzymic browning (Maillard) reactions. However, this too has been overcome by the commercial production of 'hydrogenated glucose syrups' which are recognised sweeteners in the UK and other parts of the world and which have a technological stability greater than that of sucrose in many applications (Sicard, 1982).

Ordinary glucose syrups are not classified as sweetening agents in UK legislation. They are considered to be natural food sugars and indeed, their definition by the 'Codex Alimentarius Commission' as aqueous mixtures of nutritive saccharides derived from starch, supports this categorisation. The key to their tremendous success as food ingredients lies in the flexibility of their uses. Thus glucose syrups may be designed to possess desired levels of osmotic pressure (for preservation effects), viscosity, sweetness and Maillard effects. They have made enormous contributions to crystallisation control, inhibition of foaming, lustre of food products, freezing point and fermentation control. High maltose syrups (Fullbrook, 1982) can be produced by enzymic methods containing 62–65% of the disaccharide depending on the source of the starch and these are of particular importance in the brewing and fermentation industries. At the opposite extreme, low degrees of conversion give rise to maltodextrins or 'amylodextrins' which are highly important texture-conferring ingredients.

Their nature and characterisation have been the subject of a recent study by the American Chemical Society (Friedman, 1991).

All of the different types of glucose syrup now available demand increasingly sophisticated analytical techniques to keep pace with their intrinsic variability. Vast arrays of chemical and physical procedures are now used for analysis, some of which are lengthy and expensive. It is therefore a great pleasure to note one elegantly simple method in particular, devised by one of the editors of this book (Kearsley, 1978) which obtains DE by two rapid physical determinations; refractive index, and freezing point measurement (osmotic pressure). The latter is a colligative property dependent on the total numbers of molecules. The DE is therefore determinable almost within seconds. An additional advantage of this beautifully simple procedure is that the 'equivalent DE' of a non-reducing hydrogenated glucose syrup can be elucidated in the same way.

Both the editors of this book are very experienced in the field of food carbohydrates and have a wealth of knowledge in fundamental and applied aspects of glucose syrup research. They have produced a previous volume on this subject (Dziedzic and Kearsley, 1984) and I wish all success to the launch of this important new book.

G. Birch
Professor of Food Chemistry
The University of Reading

References

Birch, G.G., Green, L.F. and Coulson, C.B. (Eds) (1970).*Glucose Syrups and Related Carbohydrates*. Elsevier Applied Science.

Dziedzic, S. and Kearsley, M.W. (Eds) (1984). *Glucose Syrups: Science and Technology*. Elsevier Applied Science.

Friedman, R.B. (Ed.) (1991). *Biotechnology of Amylodextrin Oligosaccharides*. ACS Symposium Series 458.

Fullbrook, P.D. (1982). Chapter in Birch, G.G. and Parker, K.J. (Eds), *Nutritive Sweeteners*. Elsevier Applied Science, 49–82.

Kearsley, M.W. (1978). *J. Assoc. Public Anal.* **16**, 85.

Sicard, P.J. (1982). Chapter in Birch, G.G. and Parker, K.J. (Eds), *Nutritive Sweeteners*. Elsevier Applied Science, 145–170.

Contributors

G.G. Birch	Department of Food Science and Technology, University of Reading, Whiteknights, Reading, Berkshire RG6 2AP, UK
S.Z. Dziedzic	British Sugar plc, PO Box 26, Oundle Road, Peterborough PE2 9QU, UK
P.A. Gouy	Roquette Frères, F-62136 Lestrem, France
E.B. Jackson	177 Chester Road, Hazel Grove, Stockport SK7 6EN, UK
M.W. Kearsley	British Sugar plc, PO Box 26, Oundle Road, Peterborough PE2 9QU, UK
J.F. Kennedy	Birmingham Carbohydrate and Protein Technology Group, Research Laboratory for Chemistry of Bioactive Carbohydrates and Proteins, School of Chemistry, The University of Birmingham, Edgbaston, Birmingham B15 2TT, UK
C.J. Knill	Birmingham Carbohydrate and Protein Technology Group, Research Laboratory for Chemistry of Bioactive Carbohydrates and Proteins, School of Chemistry, The University of Birmingham, Edgbaston, Birmingham B15 2TT, UK
Y. Le Bot	Roquette Frères, F-62136 Lestrem, France
G.A. Mitchell	Eridania Béghin-Say s.a./n.v., Havenstraat 84, B-1800 Vilvoorde, Belgium
T. O'Rourke	Courage Brewing Ltd., The Berkshire Brewery, Imperial Way, Reading, Berkshire RG2 0PN, UK
H.S. Olsen	Enzyme Development & Application, Novo Nordisk A/S, Novo Alle, DK-2880 Bagsvaerd, Denmark
D.S.J. Spruyt	Eridania Béghin-Say s.a./n.v., Havenstraat 84, B-1800 Vilvoorde, Belgium
D.M. Storey	Department of Biological Sciences, University of Salford, Salford M5 4WT, UK

D.W. Taylor	Chembiotech Ltd, Institute of Research and Development, Birmingham University Research Park, Vincent Drive, Edgbaston, Birmingham B15 2SQ, UK
M.J.P. Vanderbist	Eridania Béghin-Say s.a./n.v., Havenstraat 84, B-1800 Vilvoorde, Belgium
J.A. Wilson	Cerestar UK Ltd, Trafford Park, Manchester M17 1PA, UK
D. Wood	ABR Foods, Sallow Road, Weldon Industrial Estate, Corby, Northants NN17 5JX, UK
A. Zumbé	Cadbury Schweppes plc, PO Box 2138, Franklyn House, Bournville Lane, Birmingham B30 2NB, UK

Contents

1 The technology of starch production

S.Z. DZIEDZIC and M.W. KEARSLEY

1.1 Introduction

1.1.1 An overview

Both animals and plants lay down energy reserves for periods of dormancy (hibernation), germination, growth and propagation, and for survival under adverse conditions. In man, for example, the short term reserve energy source is glycogen, a short chain glucose polymer which is stored in the liver and the long term energy source is fat which is stored in the adipose tissue. In plants the reserve energy source may be either carbohydrate or oil (or fat). Examples include oil in olives and sucrose in sugar beet and cane. Most commonly however, the energy reserve in plants is starch.

Glucose (dextrose) is synthesised in the chloroplasts of cells from carbon dioxide and water in the presence of chlorophyll and sunlight. The process is commonly referred to as photosynthesis. The sugar is then transported, mainly at night, to the reserve organs of the plant where it is laid down in the polymeric form. The process is very sophisticated and may also play a secondary role in, for example, potatoes where adverse temperature conditions are counteracted by starch breaking down to the commensurate glucose. Such a breakdown decreases the average molecular weight of the cell solutes and as a consequence lowers the freezing point, thus allowing for an inherent protection mechanism to exist. Such systems have evolved where the natural habitat of the plant may have been subject to wide day and night temperature variations.

Starch is present in most tissues. It is layed down in both the granular and amorphous form in defined cells. The starch, its makeup and where it is stored will be plant specific and is as diverse as seeds (maize, wheat, sorghum, rice), tubers (potato), roots (tapioca, sweet potato, arrowroot), pith (sago palm) and fruits (banana) themselves.

As mentioned above, sources of starch are numerous and in theory any of these sources could be utilised in an industrial process. In practice, however, only a limited number are used commercially for starch production owing to the viability of the processes involved. The most important of these starch sources are maize (corn), wheat, potato and tapioca (cassava, manioc) although reference will be made in this chapter to others as appropriate.

Whatever the starch source, its use will be dependent on the economics of production. This all embracing objective will be subject to geographical location, climate, sophistication of plant, whether production is seasonal or continuous (raw material storage and transport), political considerations (financial incentives) and other factors.

Various starches have specific properties in terms of their own functionality and end use capabilities. Under 'other factors' what must be considered is whether the starch source is for the sole purpose of glucose syrup manufacture or whether this is a parallel process to starch manufacture for other purposes. These end purposes may include the specific use of the starch in food, paper and board, textiles or its ability to be modified chemically and physically to give materials with a defined functionality and use, again in the various industries mentioned above. Such needs will often take priority and will determine the starch source. Wheat starch, it could be argued, is a waste product from the manufacture of gluten. Its use, especially when the gluten price is high is therefore very desirable compared with dent maize which may be specifically grown for glucose syrup production.

Starch from some sources may additionally be considered too valuable for conversion to glucose syrup owing to the combination of its unique functional properties in native or modified form and demand exceeding supply, e.g. waxy maize starch.

1.2 Glucose syrups

1.2.1 An overview

Glucose syrups, glucose or corn syrups as they are often referred to in the USA are somewhat misleading terms as they do not describe the products correctly. Commercial products now available may in fact contain very little glucose, less than 1% in some cases, and are not necessarily manufactured from corn. Furthermore the term 'glucose' when referred to will be interpreted by many as the monomer α-D-glucose (dextrose). Whilst from a strict chemical point of view the monomer is referred to as glucose, in the glucose syrup industry the term glucose refers to glucose syrup and the monomer is called dextrose and exists as dextrose monohydrate or anhydrous dextrose.

Glucose syrups are in fact the hydrolysis products of starch and consist of some glucose monomer and varying quantities of dimer, oligosaccharides and polysaccharides, dependent on the glucose syrup in question and its process of manufacture. Other sugars may also be present as byproducts from the method of manufacture. These sugars, known as reversion products, are nevertheless important since although present in small amounts they may be the cause of undesirable off-flavours. They are commonly associated with acid hydrolysis manufacture.

Glucose syrups are in effect mixtures of anhydroglucose units, molecular weight 162 Da, formed from the condensation of two glucose units. The Codex Alimentarius defines glucose syrups as "a purified concentrated aqueous solution of nutritive saccharides obtained from starch". The EC defines glucose syrup as a concentrated aqueous solution of D(+)-glucose, maltose and other polymers of D(+)-glucose obtained by the controlled partial hydrolysis of edible starch.

The 'discovery' of glucose syrup was made during the Napoleonic wars when Europe was suffering from a naval blockade by the British. A lack of sugar from the West Indies inspired many a chemist to look for ways of synthesising a sweet alternative. A breakthrough came in 1811 when Kirchoff, a German chemist, showed that by boiling wheat starch with dilute sulphuric acid a sweet syrup could be obtained. Within a matter of a few years the whole process had achieved a considerable gearing. Potato starch was used as the starch source and sulphuric acid was replaced by hydrochloric acid and indirect heating of the reaction vessel was common place. Plants were built all over central and western Europe to capitalise on the new technology. The process also rapidly spread to the USA from where the glucose syrup industry based on maize developed. In these early days the industry developed at a faster rate in the USA and after the Second World War Europe had to buy a great deal of technology from the USA for the manufacture of starch and glucose syrup to compete on a world basis.

Glucose syrups are by convention referred to as those products with a dextrose equivalent (DE) of above 20 DE and below 80 DE. Below 20 DE the products are known as maltodextrins and above 80 DE as hydrolysates or hydrol. Within this definiton, the starch source is not important in terms of the final properties of the glucose syrup (all other things being equal). Below 20 DE the maltodextrins may take on a character dependent on the starch source, and this will be more apparent the lower the DE. This is as a direct result of the particular amylose/amylopectin makeup of the starch granule, the starch specificity. For example a wheat-derived maltodextrin of 12 DE may retrograde in solution giving a cloudy product after some time whereas a waxy maize-derived maltodextrin will not. At the lower DE levels the differences are so pronounced that different gels are formed with totally different functionalities and end uses, for example as fat substitutes. On the other hand the properties of the hydrolysates are unrelated to their source starch. This may be of fundamental importance, however, with respect to minor impurities which may exist in the starch, such as pentosans and proteins, and which may lead to unacceptable problems when the end use is pharmaceutical infusion or a chemical process which involves boiling the syrup at high concentrations.

The original process of heating starch with a mineral acid in a batch process is still in essence a valuable tool in the manufacture of glucose syrups. The batch system has however, evolved into a routine continuous process. Although the process of hydrolysis using an acid is random in nature, very good reproducibility can be achieved in terms of the product carbohydrate spectrum providing the same processing conditions are employed. In a continuous system, at the same starch dry matter, three parameters may be changed in the process to regulate the composition of the finished product, temperature (pressure), residence time and acid concentration. Modern processes will usually vary residence time (throughput speed) to obtain a constant DE and constant composition product. An upper limit of about 63 DE is imposed when using acid hydrolysis owing to the production of reversion products at higher DE levels which give the syrup a bitter taste and dark colour. Even a 63 DE syrup will require extensive cleaning up with carbon and ion exchange resins to make it water white and purely sweet. Reversion products arise through non-specific reactions between the acid and the glucose units themselves.

The development of the enzyme industry led to a quantum leap in terms of glucose syrup manufacture. Various combination methods including acid/enzyme and enzyme/enzyme are now routinely used to manufacture a range of 'tailor-made' syrups for specific purposes, and these are discussed in detail in the following chapters.

Regardless of the starch, glucose syrups within our definition above will be made to a similar carbohydrate profile and end use ability, whatever the starch source for the various methods of manufacture. This is providing the feedstock is a reasonably pure starch source. Other differences which may have occurred previously and which may have hindered or promoted a specific end-use application have now been largely overcome by improved post-hydrolysis process technology, e.g. protein removal, carbon refining and ion exchange techniques.

1.2.2 Protein removal

Starch granules naturally contain protein in their structure. The protein content varies depending on the starch source (see Table 1.2) and is seen as undesirable since it gives rise to browning reactions both during the hydrolysis process to produce the glucose syrup and on storage of the finished syrups themselves. In some starch sources the protein content may be particularly high, e.g. wheat starch and this may lead to processing difficulties. This proteinaceous matter is released during the hydrolysis process and is usually removed on a rotary vacuum filter. The filter is charged with a diatomaceous filter-aid on a batch process basis. The syrup is then drawn via a feed trough through the rotating filter. A fine water mist is usually sprayed onto the filter-aid to aid filtration and a knife edge is used to remove the spent filter-aid and protein. This avoids blocking of the filter and consequent reduced filtration rates since a new filter surface is constantly presented to the unrefined syrup. Pentosans in the wheat starch often hinder this process.

1.2.3 Carbon refining

Carbon treatment was the traditional way of cleaning up glucose syrups and is still an invaluable stage in the process. The carbon is made by activating bone charcoal and it has the ability of removing off-flavours and colours formed during the manufacturing process. Traditionally, powdered charcoal was used. This was added to the vessel containing the syrup and stirred at high temperature. The charcoal would then be removed using Candel filters. The process has evolved to make use of a granular form of charcoal in a counter-current continuous process. The charcoal is held in large vertical columns and the syrup flows from bottom to top. At regular intervals a portion of the spent carbon is removed from the bottom and fresh material is added on top. Granular carbon may be reactivated by heating in a furnace although care must be taken not to burn the charcoal. The system is expensive to implement and not as effective as using the powdered form, presumably due to surface contact area and time.

1.2.4 Ion exchange chromatography

The process is now a common feature in many plants for the manufacture of high purity syrups. The main purpose of the process is to lower the ash content from what could be 0.35–0.5% in a traditional acid process to approximately 0.01%. Mineral ions catalyse browning reactions and their removal produces a much more stable syrup. An added advantage of ion exchange treatment is that many coloured bodies formed by Maillard reactions during syrup manufacture are themselves charged and these are additionally removed. These charged bodies are precursors of the typical brown Maillard pigments and their removal gives the syrup prolonged shelf life.

Ion exchange employs two separate columns, a cation exchange and an anion exchange resin in series. The resistivity of the glucose syrups is measured and this can be calibrated against a predetermined mineral content. This also enables continuous monitoring of the column quality. An increase in conductivity signals the need to change the column and to reactivate the resin.

Some users argue for the continued use of the higher mineral content syrups due to their better taste qualities and specific uses, for example in certain confectionery, soft drink and preserve applications. The manufacturers counter this with claims of better stability, uniformity of product coupled with the ability to add back minerals to a recipe to achieve a consistent product. On balance the manufacturers are winning the argument especially when taking into account the fact that most bulk deliveries and storage occur at about 50°C, a temperature where colour development could be significant on prolonged storage.

A final evaporation of the syrup by a series of rising-and/or falling-film evaporators is employed. The final solids content is a compromise between stability and ease of handling and will normally range from 72% to 81% solids.

1.2.5 Conclusion

Taking into account the above process manufacturing parameters and within the definition of what is termed glucose syrup as opposed to maltodextrin or hydrolysate, we can conclude that syrups manufactured from various sources by the same processes to the same DE will have comparable physical and chemical properties.

1.3 Starch

1.3.1 Origins of use

The origins of starch as an industrial material date back to Egyptian times when it was used in the manufacture of papyrus parchment and starch-based adhesive. Various other references exist to its use in Roman times as a whitener and stiffener for textiles. In Europe the starch industry was mainly based on potato and evolved on the back of the textile industry where its use as a sizing agent was its main outlet. Post war Europe realised that the seasonal nature of potato starch manufacture could be counter-balanced through the continuous milling of a maize crop. This in essence was the beginning of the diversification to the various starch sources which we encounter today.

1.3.2 Occurrence and appearance

As mentioned in the introductory section, starch is layed down in the tissues of many plants. The commercial realities of the starch recovery process however limit the industrial sources mainly to wheat, maize, potato and tapioca. The starch is found in granules whose shape and size are specific to the starch source. The starch source can be relatively easily identified from the extracted starch by viewing the granules under a medium powered microscope. An iodine-based stain is commonly used for clarity and will also give further clues as to the amylose/amylopectin content of the granule. The granules vary from 2 to 150 μm in size, and are also characterised by specific shapes which range from round to polygonal and truncated. Figure 1.1 shows a range of the more common starches and Table 1.1 lists their particle size and structure type.

(a)

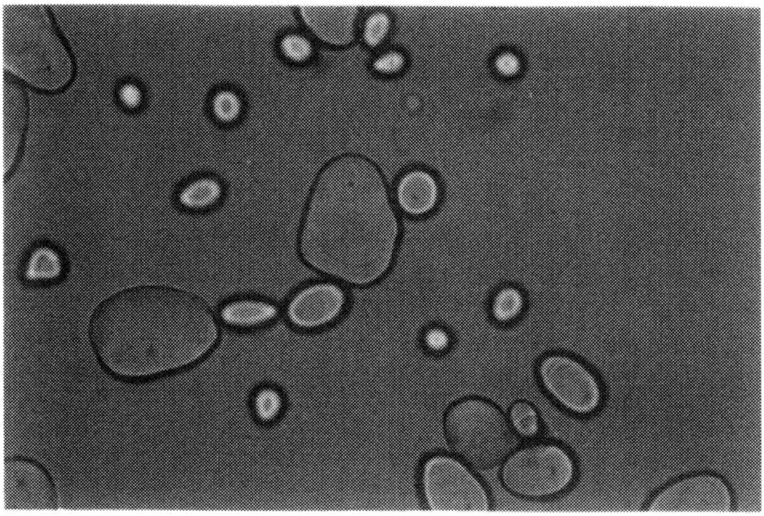

(b)

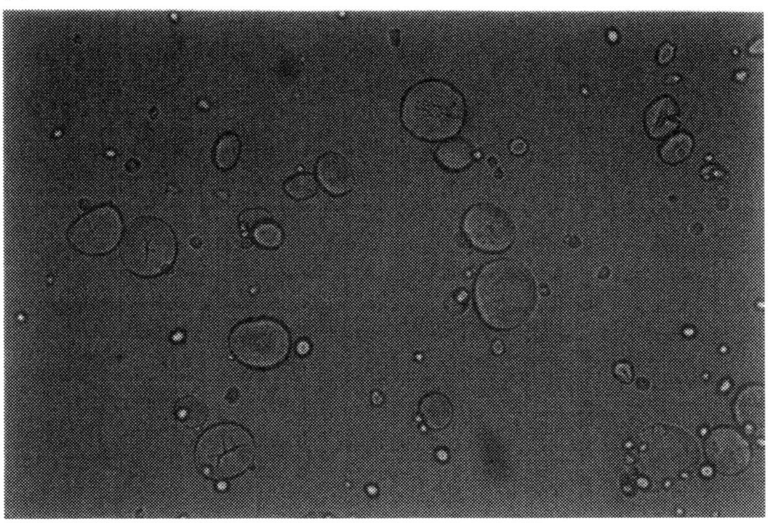

Figure 1.1 Shape and size ranges of granules of common starches ×260. (a) waxy maize starch, (b) potato starch, (c) tapioca starch, (d) wheat starch.

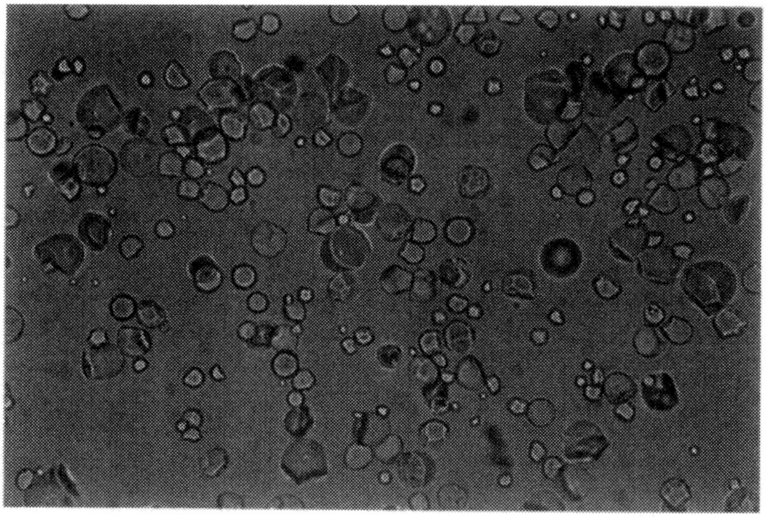

(c)

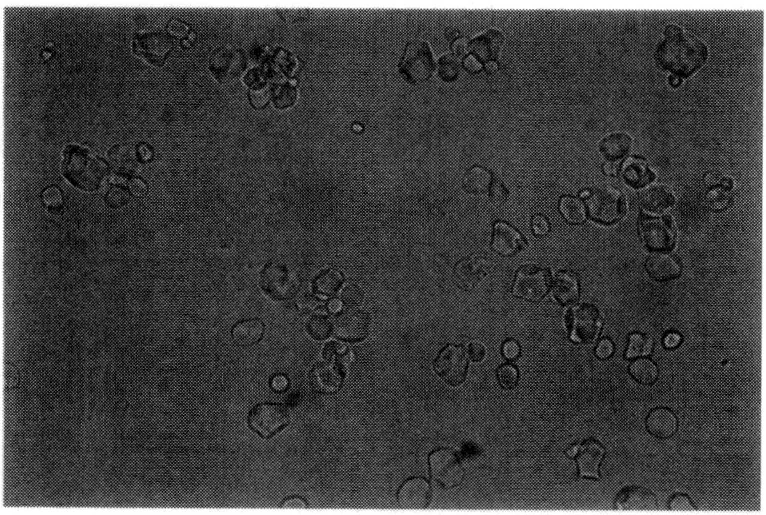

(d)

Figure 1.1 *continued*

Table 1.1 Characteristic size and shape of common starches

Starch type	Diameter size range (μm)	Shape
Maize	5–30	Round, polyhedral
Wheat	2–40	Round
Potato	50–150	Ovoid
Tapioca	4–30	Kettle-shape

1.3.3 Raw material composition

The composition of the raw material will be dependent on various factors, such as variety (of potato, maize etc.) and climatic conditions during growth and time of harvest. Storage conditions prior to processing may also influence minor component levels such as sugars. In the manufacture of starch for the production of glucose syrups, these will have little consequence except for perhaps pentosans and proteins which may lead to certain problems in processing and end use. Otherwise, adverse conditions of growth, harvest or storage of the raw material will only lead to reduced overall yields. Modern starch plants are designed to process a range of raw material qualities by making minor changes to the processing conditions. This is certainly not true for the manufacture of starch for its use *per se* or for the manufacture of products such as potato granules and flake, where the quality of the raw material and the fundamental composition of the starch may have pronounced effects on its efficacy in a given final manufacturing process, for example, in the manufacture of snack pellets. Table 1.2 lists the common starch sources and their typical composition.

Some interesting conclusions may be drawn from this basic table which gives some underlying criteria for raw material suitability for glucose production.

The starch content of the four raw materials on a dry weight basis is quite similar. However, when comparing the moisture contents of potato and tapioca starch (tuber starches) to those of wheat and maize starch (cereal starches) we can see a major difference. The tubers have considerably more moisture and the economics of production are therefore very different based on this fundamental difference. In Europe, for potatoes, a 30 mile radius is typically the limit of the harvest region owing to the unacceptable cost of transporting what is in effect 80% water. Potatoes are harvested on a campaign basis which usually lasts between August and January. The campaign may be extended providing no adverse weather conditions occur or if there is the possibility of adequate storage, usually underground. If the potato crop is subject to frost damage and subsequent spoilage it should be processed as soon as possible, being transformed into the stable starch. This may be further processed into glucose syrup or stored as the dried starch until required. The economics of storing the

Table 1.2 Typical percentage composition of common raw materials

Source	Starch	Moisture	Protein	Fat	Fibre
Maize	61	16	9	4	2
Potato	19	75	2	0.2	1.6
Wheat	65	14	13	2	3.5
Tapioca	30	63	3	0.3	5.5

starch and its subsequent processing to glucose syrup are however not favourable because of the increased costs of producing dry starch.

Tapioca also has a high moisture content and is subject to even more regional growing restrictions around the starch plant than the potato since in the areas where it is grown there is usually a poor infrastructure which is often coupled to inadequate storage facilities. Furthermore, tapioca in the sub-tropical climate where it is grown is very susceptible to spoilage and must be processed almost immediately on arrival at the plant. If processing does not commence within the first 24 h then it is necessary to dry the tuber to prevent spoilage.

The production of cereal starch on the other hand is well adapted to a continuous yearly production process. Both wheat and maize can be grown away from the immediate vicinity of the plant where they can be dried and stored quite favourably until required for processing. This gives the process a great deal of flexibility and further allows for material to be bought on the world spot markets to supplement production needs as required.

The protein content variation between tubers and cereals is quite significant. This protein content is usually expressed as Kjeldahl nitrogen × 6.25 and will include all nitrogen-containing compounds, such as any peptides, amides and amino acids that may be present. The starch recovery processes which will be discussed later are adapted to these botanical variations and maximise the yield of starch in its purest form.

The lipid fraction consists of triglyceride esters and free fatty acids (mainly palmitic, oleic and linoleic acids) together with some complex lipids (mainly phospholipids). The lipid content is substantially higher in cereals compared with tubers. The oil fraction occurs mainly in the germ of the cereal and is removed during the early stages of the starch manufacturing process.

The fibre content, as would be expected, is also higher in wheat and maize and primarily comes from the hull component. As with the germ it is mainly removed during the first stages of starch manufacture.

1.3.4 Starch composition

As previously discussed, starch consists of granules which are diverse in shape and size and are specific to their botanical source. The starch granule consists of two glucose polymer structures, namely amylose and amylopectin, and also contains moisture, lipids, proteins and mineral ions either present discretely in the matrix of the starch granule or as a specific complex with the starch polymer. Table 1.3 gives typical analyses for the common starches.

1.3.4.1 Moisture. The moisture content of a starch will vary depending on the surrounding conditions in which it is placed. Moisture contents are

Table 1.3 Typical percentage composition of common starches (dry solids basis)

Starch	Moisture	Lipid	Protein	Ash
Maize	13	0.7	0.15	0.12
Potato	19	0.08	0.06	0.1
Wheat	14	0.8	0.3	0.15
Tapioca	13	0.2	0.2	0.2

usually quoted at 65% relative humidity and 20°C, typical ambient conditions. Under these conditions the moisture content of maize, wheat and tapioca is usually quoted at between 13–14%, and that of potato between 18–20%. Starches are usually supplied commercially at these equilibrium moisture contents to limit the risk of any dust explosions. The ability of starch to gain and lose large quantities of moisture is utilised in confectionery in starch-depositing processes.

1.3.4.2 Lipids. As can be seen in Table 1.3, higher lipid levels are found in cereal than tuber/root starches. Lipids are of importance in the starch *per se* since they can form complexes with the starch polymers and make a significant contribution to the behavioural characteristics of the starch. Lipids can also impart off-flavours to a native or modified starch. These off-flavours can be significantly reduced by extensive washing of the starch in either native or modified form. Lipids have little significance as far as glucose syrup production is concerned.

1.3.4.3 Proteins. Both maize and wheat starch show a significantly higher residual protein level when compared to tapioca and potato starch. The proteinaceous matter is more important in a food starch application than in glucose syrup manufacture due to off-taints that may develop. These are referred to as specific 'cereal flavour notes' and are important in delicate tasting foods, for example instant desserts where a bland tasting starch is a priority. Since the proteins and peptides have surface activity, of far more significance is the problem of foaming caused by these residues. If not removed they can contribute to unacceptable levels of browning and foaming during concentration or during subsequent use. These residual proteins can be reduced to acceptable levels through the correct use of enzymes and refining techniques. Furthermore, wheat starch consists of two distinct granular fractions, a small spherical fraction, 2–10 μm in diameter, and a second fraction consisting of large lenticular granules, 20–40 μm in diameter. Although the large granules only account for about 10% by number, they contribute about 90% of the weight of the starch. They also contain a relatively low level of protein compared with the small granules; 0.1% compared with 1.5%. Several manufacturers now divide the starch stream into the two granular fractions to take account of this.

1.3.4.4 Ash. All starches contain mineral ions. The cations are mainly sodium, potassium and magnesium but calcium, aluminium and iron are also found. Potato starch shows a significantly higher ash content than the other three major products. This higher level is a direct result of the high phosphate levels found in potato.

1.3.4.5 Amylose and amylopectin. Starch is not a homogeneous substance but is made up of two related but structurally different polymers, amylose and amylopectin. The starch granule has a biological nucleus around which the granule is built up. This is known as the hilum and may be concentric or eccentric depending on the botanical source. Three enzymes are thought to be involved in the process of granule development, phosphorylase, starch synthetase and branching enzyme. Table 1.4 gives the amylose and amylopectin content of different starches.

The amylose content varies between 14% and 27% depending on the starch source although different varieties of the same source, e.g. maize, will have different amylose:amylopectin ratios. Specific starch varieties are now commercially available containing up to 99% amylopectin (the waxy starches) or up to 70% amylose (the high amylose starches). These starches have specific properties, are subject to growing restrictions and command a premium price. They would not normally be used in the manufacture of glucose syrups except perhaps as part of a reprocessing operation.

The basic building block of both amylose and amylopectin is the glucose monomer linked through the α-1,4 glycosidic bond (Figure 1.2). Another common way of describing the structures is that of polymeric condensation products of anhydroglucose units. During the development of the starch granule there is a slight increase in the amylose to amylopectin ratio, with a commensurate increase in the molecular weights of both the polymers. The cereal starches have a higher amylose content than the tuber starches.

An indicative degree of polymerisation (DP) of amylose is 10^2–10^3, with tuber starches showing in general a higher DP than the cereal starches. Amylopectin is usually considered as having a DP of an order of magnitude higher than amylose.

Various studies have shown that amylose is a straight chain glucose polymer with a conformational helical structure, although there is evidence

Table 1.4 Typical amylose and amylopectin content of starches

Starch	Amylose (%)	Amylopectin (%)
Maize	27	73
Potato	21	79
Wheat	27	73
Tapioca	14	86

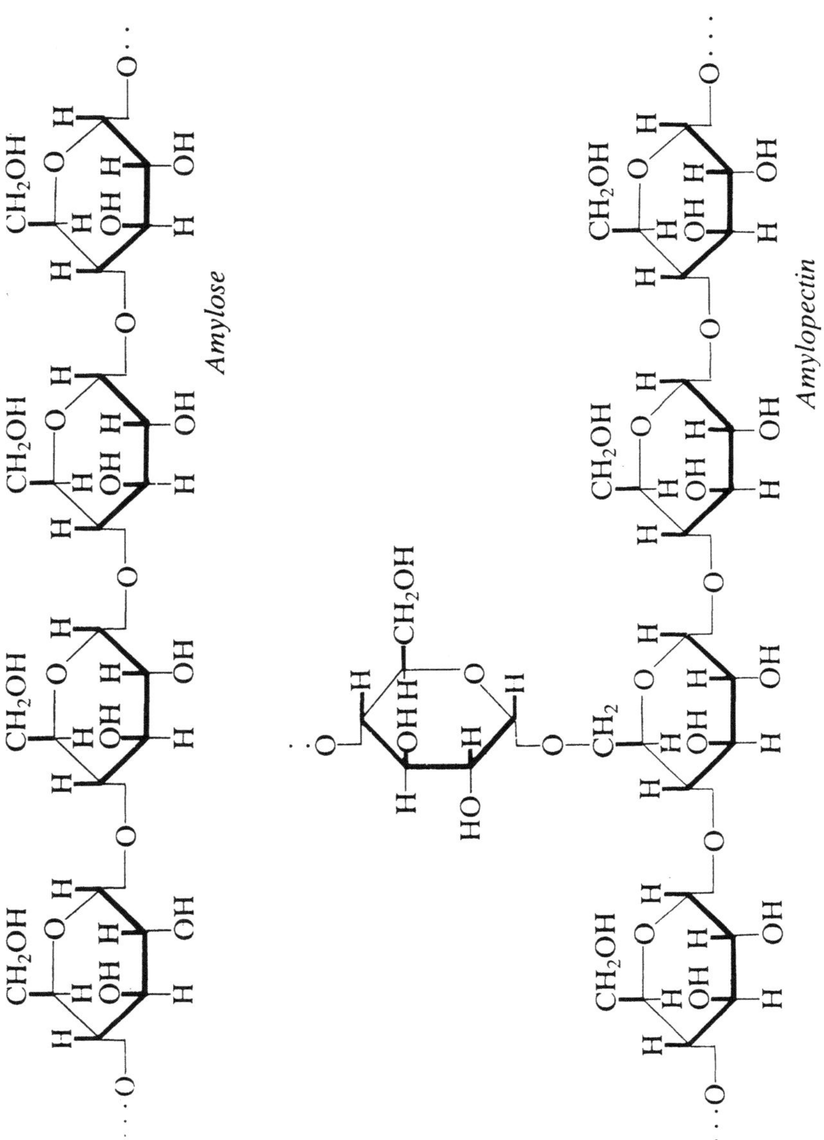

Figure 1.2 Amylose and amylopectin structure.

to suggest that a very small amount of branching may be present. Amylose has the ability to form inclusion complexes with iodine. The iodine molecule is held within a clathrate structure, with one iodine molecule per turn of the helix (six glucose units). The amylose–iodine complex has a characteristic blue colour. The colour and intensity observed are related to chain length (Table 1.5).

Amylopectin is a highly complex branched molecule. The backbone of the molecule is still the (1–4) linked anhydroglucose unit but with branching points occurring at the carbon-6 primary alcohol group (Figure 1.2) every 10–12 glucose units. There is also occasional branching at carbon-3. The side chains off the main backbone are 20–30 glucose units in length. The molecular structure is not uniform and has areas of high branching and also of relative regularity. Starch granules consist of both crystalline and amorphous regions. Amylopectin occurs in both the amorphous and crystalline areas of the granule but is considered to be solely responsible for the crystallinity of starch. Amylose is associated with the amorphous regions of the granule.

When viewed under polarised light, starch granules show a birefringence pattern often referred to as a maltese cross as shown in Figure 1.3 for various starches. The centre of the cross for a specific starch granule will be concentric or eccentric and will correspond to the hilum of the starch granule. These light diffraction patterns infer an eccentric or concentric radial organisation to starch, originating at the hilum.

Different starches show birefringence patterns of varying strength indicating differing degrees of crystallinity in individual starches. This is shown in Figure 1.3.

Both intra- and inter-molecular hydrogen bonding are responsible for the structural integrity of the starch granule and also account for its insolubility in water. When a starch suspension is heated in water, heat overcomes the hydrogen bonding forces which hold the granule together allowing swelling or gelatinisation of the starch granule to occur. This energy input is the first stage of both the acid and enzyme-catalysed hydrolysis reactions since acid and enzymes have little hydrolysing effect on the ungelatinised starch.

Table 1.5 Typical colour of iodine/starch complex with respect to starch chain length

Degree of polymerisation	Colour of starch/iodine complex
<12	Yellow
<20	Red
<30	Purple
30–40 DP	Mauve/blue
>40 DP	Blue

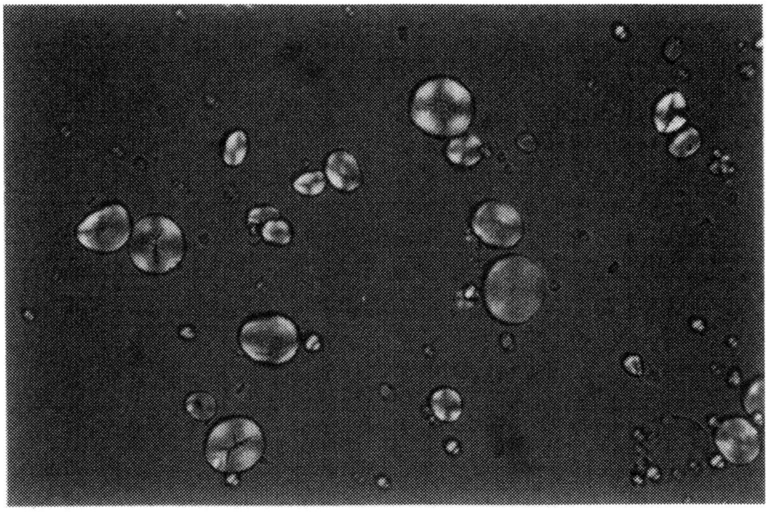

(a)

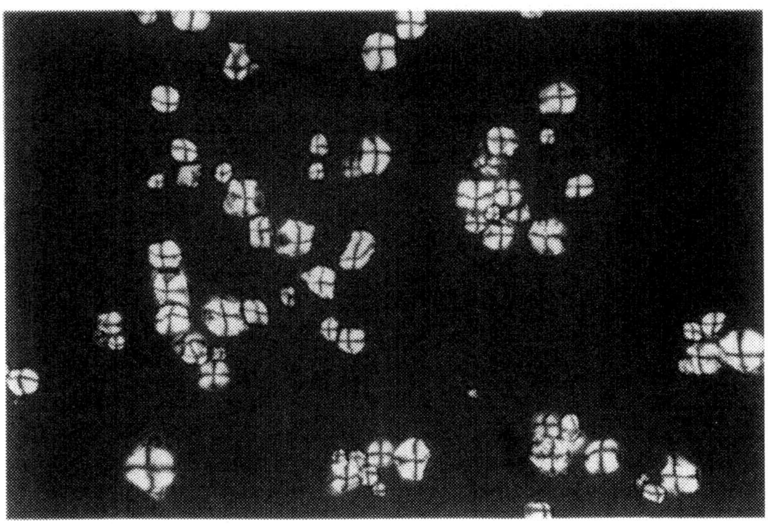

(b)

Figure 1.3 Varying birefringence patterns for different starches under plane-polarised light ×260. (a) waxy maize starch, (b) potato starch, (c) tapioca starch, (d) wheat starch.

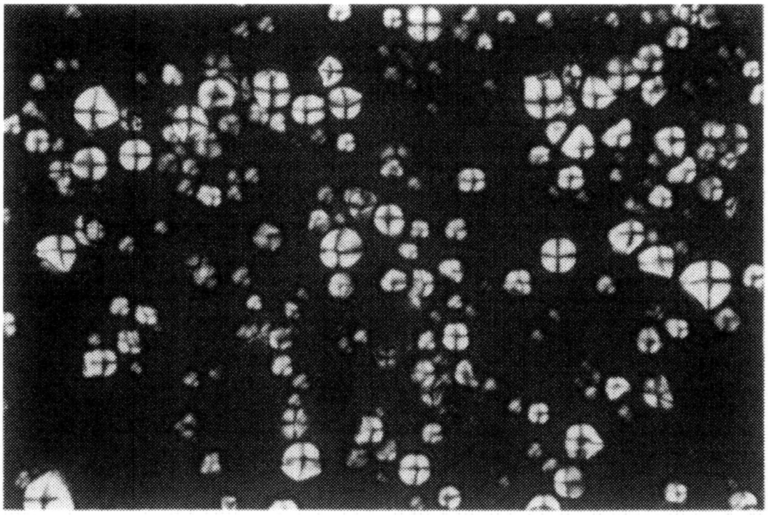

(c)

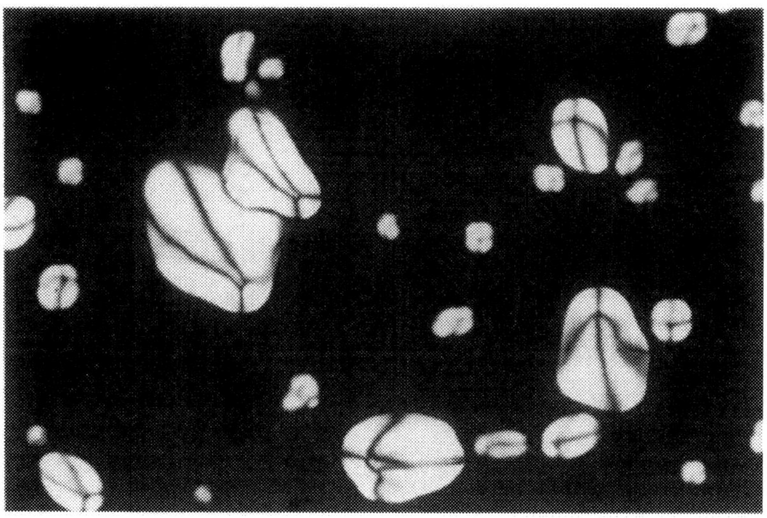

(d)

Figure 1.3 *continued*

1.4 Industrial starch recovery

1.4.1 Introduction

Methods of starch recovery vary greatly depending on country, age of plant and starting material. They all however follow the simple principle of extracting the starch in the highest possible yield. Methods often use repeated wet grinding techniques coupled with cyclone separators or conventional centrifuges and are based on the water insolubility of starch and variations in density and size of the different granule components to obtain a pure starch product. Almost all commercial starch manufacture is now carried out using continuous processes.

In the recovery of starch from whatever source it is important not to neglect the recovery of the other grain or tuber constituents, e.g. bran(fibre), protein and fat. It is usual practice to recover as much of these as possible for sale into a range of industries, e.g. animal feed and fermentation, so that nothing (or very little) goes to waste. This ensures low effluent treatment costs and increased revenue for the starch producer.

1.4.2 Maize starch

Maize starch is recovered from maize nibs by a wet milling process. In Europe most plants use home grown dent corn varieties. The schematic outline of the process is shown in Figure 1.4.

Dry milling processes have also been developed but the high oil content of maize (in the germ) lends itself more readily to the wet process.

1.4.2.1 Maize delivery, storage and cleaning. Maize is delivered to starch plants by water, rail or road. Deliveries are normally made on a daily basis but a 10–14 day buffer supply is normally maintained.

Initial cleaning of the grain is carried out by first tipping the maize through coarse security sieves to remove any larger contaminants that may be present. The grain is then pneumatically transferred to a series of vibratory sieves and winnowing tower to remove both light and heavy contamination. Broken and damaged nibs must be removed at this stage as they are unsuitable for processing. These can be sold on directly or blended with other byproducts which are usually compounded for animal feed purposes. The clean maize nibs are then pneumatically transferred to holding silos to await processing.

1.4.2.2 Steeping. The first processing stage is steeping or soaking the grain in water. Steeping has two main purposes. First, it softens the grain to allow milling without shattering and secondly it brings the moisture content to the optimum for milling to be carried out and permit the gentle

Maize Receipt	——	quality analysis
↓		
Storage		
↓		
Cleaning	——	broken maize/debris to animal feed
↓		
Steeping	——	corn steep liquor – fermentation
↓		
Coarse Milling		
↓		
Degermination	——	germ-oil extraction
↓		
Second Milling		
↓		
Degermination		
↓		
Fine Milling		
↓		
Fibre Separation	——	animal feed
↓		
Desanding		
↓		
Gluten Separation	——	animal feed/fermentation
↓		
Starch Refining		
↓		
Maize Starch	——	glucose syrups

Figure 1.4 Maize starch production.

separation of the grain into its component parts. These can be placed in four constituent groups; the outer hull (bran or fibre), the germ, the maize gluten and the starch.

Maize is typically soaked for 30–50 h in stainless steel tanks holding 50–60 tons of product at 45–60°C using a counter-current process. High temperature/pressure processes are also used to reduce steeping times and volumes of steep water produced. The water contains added sodium metabisulphite as 0.1–0.2% sulpur dioxide and is inoculated with *lactobacillus*. Sodium metabisulphite is added for several reasons. It decreases the pH, it limits unwanted fermentation from putrefactive microorganisms and it promotes *lactobacillus* growth. This last point would appear to be of paramount importance in maximising starch yield. It also acts in a disruptive way on the protein matrix in the endosperm allowing the maximum release of starch during the milling stage.

During the steeping process, the maize grains will absorb 50% of their own weight of water and lose some 6% solids. This solid matter consists of soluble low molecular weight sugars, proteins, low molecular weight peptides, amino acids and minerals. The outflow steep stream is concentrated and sold as corn steep liquor to the fermentation industry or spray dried onto maize gluten and sold as the solid for the same purpose.

Although the steeping process is discussed as a routine part of the wet milling process it must be pointed out that this stage is of considerable importance in terms of maximising the end yield of the starch and its physicochemical characteristics and is also important with respect to the quality of the corn steep liquor itself. Temperature/time changes during steeping will, for example, change the composition of the corn steep liquor. Since corn steep liquor is used in fermentation for secondary metabolite harvesting, qualities which lead to enhanced yields and optimum composition command a premium price.

1.4.2.3 Degermination. Degermination is the first stage of the separation process and is carried out by passing the soaked maize through primary attrition or pin mills. The milling surfaces have a series of interlocking pins which in fact do not mill as the name infers but rather gently tear the grains apart removing the hull and releasing the germ from the endosperm. The germ which contains most of the oil can be easily separated from the fibre, starch and protein components on the basis of its low specific gravity. Originally germ separation was accomplished in flotation tanks but now the process is carried out bypassing the process stream through multistage hydrocyclones. The germ fraction is removed as the overflow stream where it is washed several times to remove adherent starch and protein. The wash stream is added back to the process stream which then passes to a second milling stage. This removes any remaining germ from the maize kernel. A similar process of hydrocyclone separation and washing follows.

The separated germ is then dewatered and dried and may be sold for animal feed use or more commonly is deoiled on site before being used for the same purpose. The dry germ contains 50–60% oil and the extracted and refined product is widely used domestically and commercially.

1.4.2.4 Fine milling. The underflow stream, post germ separation, then passes through a fine milling stage which produces a finely ground mixture of starch, gluten and fibre. This process must be very carefully controlled to ensure maximum size reduction of the protein and starch (without damaging the granule structure) but yet leave the fibre intact as far as is possible. This facilitates separation of the fibre from the protein and starch and ensures minimum contamination of the starch stream with small fibre particles. The fibre fraction is sieved out using multistage bow sieves. It is then dewatered and dried and used for animal feed. This leaves a main stream consisting principally of starch and gluten.

1.4.2.5 Gluten separation. The gluten–starch stream is then usually passed through a desanding hydrocyclone to remove grit and other heavy particles. The primary separation of the starch from the gluten is carried out using continuous high speed centrifuges which separate the lighter gluten fraction from the heavier starch. Thorough washing of the gluten to recover as much starch as possible is essential to give good starch yields from the process. The protein is used for animal feed as it lacks functionality and has a poor taste.

1.4.2.6 Starch refining. To obtain a high quality pure starch, the final stage of the process involves the washing of the starch to remove any remaining fibre, gluten and solubles via a counter-current, multistage hydrocyclone arrangement. The resultant starch stream is usually about 40% dry solids with a protein content less than 0.15% (dry basis). The starch from this stage can be fed directly to the first step of the glucose syrup conversion process or alternatively dewatered and flash dried to give maize or corn starch.

1.4.3 Wheat starch

Unlike the wet milling process for maize, wheat starch is manufactured not from whole wheat but from wheat flour obtained from a typical wheat milling process. The starch is generally not the main grain constituent of interest to the manufacturer but is rather of secondary importance after the wheat gluten. It is however an important constituent and its removal and subsequent use has a significant effect on the profitability of the wheat processor. The overall process for wheat starch production is shown in Figure 1.5.

In the flour mill the wheat is conditioned and subjected to repeated dry milling and sieving to remove the bran or fibre and germ. Oil is not usually extracted from the germ which is typically used intact for bread making. A dough is made from the flour by adding water and after a short resting period is washed extensively to separate the starch from the protein.

It should be noted that the wheat starch industry has evolved from the gluten industry. Initially wheat starch was produced as a byproduct of this process and the glucose syrup manufactured from it was often of poor colour and quality. The situation has now significantly changed with glucose syrups being manufactured from wheat starch to high specification. Furthermore owing to the high price of gluten and other considerations, wheat is now becoming a very attractive raw material for processing in Europe.

1.4.3.1 Dough preparation. Dough is usually made up in conventional Z-blade mixers of the type commonly used throughout the baking industry. Water is added to the dough at about 60% of the flour weight. After mixing, the dough is rested for about 20 min in order for the gluten matrix to develop and facilitate separation of the starch and protein.

Wheat Receipt	——	quality analysis
↓		
Flour Production	——	germ/bran – food use
↓		
Dough Preparation		
↓		
Resting		
↓		
Starch Extraction (Washing)	——	gluten – dried
↓		
Starch		
↓		
Screening/Washing		
↓		
Wheat Starch	——	glucose syrup

Figure 1.5 Wheat starch production.

1.4.3.2 Dough extraction. Post resting the dough is washed in gluten separators which are in effect continuous extractors. The separators are usually long cylindrical vessels housing multiple horizontally rotating paddle washers. The rested dough is fed via a top inlet falling to the bottom of the trough where it is moved and kneaded via the paddle action. Water is fed in at a constant rate at the bottom of the trough to maintain a level not far above the paddles. The washed out starch is held in suspension and removed at outlets at the top water margin at approximately 10% dry substance. In a continuous process the washed gluten is removed at one end of the extractor via an outlet auger.

1.4.3.3 Starch refining. The first stage of starch refining is carried out by screening the starch stream to remove residual bran and gluten. This can be done in a number of ways but is conveniently carried out by feeding the stream through what could be best described as large scale washing machines. The starch is washed out through cylindrical screens and the residual contamination of bran and gluten is taken out through a horizontal axis augar.

The starch stream is then washed counter currently through multibank hydrocyclones to produce a starch stream which can be dried or thickened by centrifugation prior to feeding onto the conversion process.

1.4.4 Potato starch

Unlike corn and wheat where material may be bought against internationally agreed quality specifications, potatoes are usually supplied by farmers local to the plant. The farmers are contracted to grow a specific acreage of potatoes and are supplied with seed potatoes of industrial quality. The harvested crop contains up to 20% starch compared with domestic potato varieties which contain about 10% starch. Supply from such a diverse group of producers means that the quality of the raw material will vary considerably, for example regarding contamination, residual soil, etc. and incoming potatoes at the factory must be closely examined.

Many statistics are quoted in the potato starch industry. These include the energy required to treat the waste which is equivalent to that required to extract the starch and in a potato starch plant half the surface area is required to process the waste (usually in a series of lagoons). It must be emphasised that environmental issues are a major problem and the plants efficiency and cost effectiveness will depend on the ability to recycle water and remove as much organic waste from the effluent stream as possible. The tonnage of water required to process one ton of potatoes must be as small as possible.

1.4.4.1 Delivery and cleaning. The overall process of potato starch production is shown in Figure 1.6.

Potatoes must be processed quickly after harvest or they will rapidly deteriorate, unlike maize and wheat which can be dried and stored almost indefinitely. A wide range of transport is used to deliver potatoes to the processing plant although road is the usual means owing to the local nature of the raw material. All deliveries are weighed and randomly sampled to assess the content of soil, stones and other debris. These contaminants are removed from the sample and weighed to form a measure of the 'impurities' in the product. The starch content of the potatoes is determined by weighing a known quantity of potatoes in air and water and calculating the starch content by reference to density tables.

After tipping the potatoes in a 'dirty' holding area they are taken into a water flume to sieves where floating vegetable matter and stones are removed. The potatoes are then washed in rotating drums were most of the dirt is removed by attrition. Many thousands of tons of soil, stones and vegetable matter are produced during a typical campaign and the weather at harvest of the potatoes is important with respect to the quantities of these components in the potatoes. The potatoes are then stored in a clean potato area ready for processing.

1.4.4.2 Rasping, extraction and refining. Potatoes are a relatively pure starch source and can be processed fairly simple. They are first milled by gravity feeding onto rotating rasping saws which reduce the potatoes to a pulp. This process must destroy the structure of the potato and free the starch from the other constituents without damaging the starch granules or producing too fine a pulp. The pulp is passed over magnets to remove any possible metal contamination from the rasping stage. Sodium meta-

Potato Receipt	——	quality analysis (starch)
↓		
Washing	——	soil/debris/stones
↓		
Milling (Rasping)		
↓		
Fruit Water Separation	——	protein recovery
↓		
Pulp Removal	——	animal feed
↓		
Desanding		
↓		
Starch		
↓		
Screening/Washing		
↓		
Potato Starch	——	glucose syrup

Figure 1.6 Potato starch production.

bisulphite is added to limit enzymic browning and metal complexing reactions which may occur. The potato pulp is then passed to a slurry tank where as much fruit water as possible is removed by decantation.

Antimicrobial agents may also be added at this stage to reduce the numbers of microorganisms in the finished product. The starch stream is purified to remove any heavy debris, fibre and soluble matter by subjecting it to repeated cycles through multibank hydrocyclones. The final step involves washing the starch counter currently with conditioned water. The starch stream may then either be dewatered to 60–65% solids on rotary vacuum filters and dried or concentrated to 30–40% solids for feeding onto conversion into glucose syrups. Potato fibre and protein may also be recovered during the process of starch manufacture. Although these constituents are not present in large quantities in potatoes, over a campaign, where many hundreds of thousands of tons of potatoes are processed significant quantities are recoverable. Typically these components are used for animal feed. The fruit water is a particularly good source of soluble protein which can be recovered by heat coagulation.

1.4.5 Tapioca starch

Tapioca is a common starch source in Brazil, Thailand, Malaysia, Indonesia and Nigeria. The processing of tapioca follows a very similar process to that for potatoes. Both crops have high water contents and have to be processed as soon as possible after harvest. The tapioca root varies widely in size between about 20–75 cm in length and is also covered with a skin that is removed prior to processing. The peeled roots are cut into thin slices prior to grinding in the rasping mill. The tapioca slurry is then subjected to a similar processing regime as potato slurry. The process is shown in Figure 1.7.

In many of the countries which use tapioca as a starch source, roots are available all year round and processing is therefore virtually non-stop. The quality of the roots and particularly their composition will vary over the year, however, with starch contents for example ranging from 13% to 25% over the processing period. Viscosity of the extracted starch also changes markedly over the campaign period.

Farmers are paid according to the starch content of the tapioca and there is obviously some reluctance among farmers to bring low starch content roots to the factory just to keep the factory running. Where the country has a wet season, actually harvesting roots may be impossible and factories have no option but to stop production of starch. A further complication with tapioca roots is that some varieties contain appreciable quantities of cyanide and care must be taken to ensure the starch is washed thoroughly to remove all traces.

Although the production techniques for tapioca starch are very similar to those for potato starch, tapioca starch production is not as well developed as potato starch. For example, the pulp removed from the roots and sent for animal feed contains up to 55% starch which is a considerable loss of revenue.

1.5 Conclusion

The technology needed to manufacture starches from a very wide range of raw materials is readily available. Economics will however in most cases dictate the actual material of choice. Cost efficiencies will similarly dictate which raw material is used for glucose syrup manufacture providing that

Tapioca Roots Receipt	——	quality analysis (starch)
↓		
Topping/Tailing	——	waste/animal feed
↓		
Washing	——	soil/debris/stones
↓		
Milling (Rasping)		
↓		
Decantation (increase solids)		
↓		
Fibre Removal	——	animal feed
↓		
Starch Concentration		
↓		
Tapioca Starch	——	glucose syrup

Figure 1.7 Tapioca starch production.

the priority use is not the starch itself. Glucose syrups can similarly be manufactured to consistent and comparable qualities from the different starch sources using existing technologies with minimal differences in the finished products.

Bibliography

Corn Refiners Association (1975) *Critical Data Tables*, Corn Refiners Association, 1001 Connecticut Ave, N.W. Washington D.C. 20036.

Corn Refiners Association (1992) *Standard Analytical Methods*, Corn Refiners Association, 1001 Connecticut Ave, N.W. Washington D.C. 20036.

J. Dobbing (ed.) (1992) *Dietary Starches and Sugars in Man*, Springer-Verlag.

J.A. Radley (1976) *Examination and Analysis of Starch and Starch Products*, Applied Science Publishers.

J.A. Radley (1976) *Industrial Uses of Starch and its Derivatives*, Applied Science Publishers.

G.G. Birch, L.F. Green and C.B. Cowson (eds) (1970) *Glucose Syrups and Related Carbohydrates*, Elsevier.

S.Z. Dziedzic and M.W. Kearsley (eds) (1984) *Glucose Syrups: Science and Technology*, Elsevier Applied Science.

W.R. Junk and H.M. Pancoast (1973) *Handbook of Sugars*, Avi Publishing.

R.L. Whistler (1964) *Methods in Carbohydrate Chemistry, Vol. IV. Starch*, Academic Press.

R.C. Hoseney (1986) *Principles of Cereal Science and Technology*, American Association of Cereal Chemists.

J.A. Radley (ed.) (1968) *Starch and Its Derivatives* 4th edn., Chapman and Hall.

R.L. Whistler and E.F. Paschall (1965) *Starch Chemistry and Technology, Vol. 1, Fundamental Aspects*, Academic Press.

G.M.A. Van Beynum and J.A. Roels (eds.) (1985) *Starch Conversion Technology*, Marcel Dekker.

F.W. Schenke and R.E. Hebeda (eds.), (1992) *Starch Hydrolysis Products*, VCH.

2 Enzymatic production of glucose syrups

H.S. OLSEN

2.1 Introduction

Industrial enzymology is an important branch of biotechnology. Enzymatic processes permit natural raw materials to be upgraded and finished. Enzymes offer alternative ways of making products previously made using conventional chemical processes.

Enzymes were used in ancient Greece for the production of cheese. Early references to this are found in Greek epic poems dating from about 800 BC. Fermentation processes for brewing, baking and the production of alcohol have been known since prehistoric times.

The development of submerged fermentation techniques has resulted in tremendous progress in the field of industrial enzymology. This technique was already in use in the early 1950s for production of bacterial amylase for the textile industry. Microbial enzymes were used in the starch industry before their use in the detergent industry. Special types of syrups that could not be made by means of conventional chemical hydrolysis were the first compounds made entirely by enzyme processes.

The actual turning point for the food industry was reached in the early 1960s when an enzyme preparation (amyloglucosidase) was launched, for the first time, by which starch could be broken down completely into glucose. Soon thereafter almost all glucose production was changed to enzymic hydrolysis from acid hydrolysis; not least because of advantages that were obvious to the producer, namely greater yields, higher degrees of purity and facilitated crystallization. The most significant event, the development of immobilized glucose isomerase, which made the industrial production of fructose syrup feasible, took place in 1973. Since then the starch conversion industry has become the second largest enzyme consuming industry, and due to the many different products made from starch, significant research efforts have been devoted to it. Modern biotechnology using the latest techniques of molecular biology and genetic engineering has been concentrated very much on this practical industry. Apart from enzyme development, a significant amount of application process research has been the necessary backup to support breakthroughs in microbiology.

Enzymes are ideal catalysts for the food industry owing to their efficiency, specific action, ability to work under mild conditions, and their high purification and standardization. Simple equipment can be used, and as moderate temperature and pH values are required for the reactions only few byproducts influencing taste and colour are formed. Furthermore enzyme reactions are easily controlled and can be stopped when the desired degree of conversion is reached.

2.1.1 World consumption of industrial enzymes in food processing

Worldwide consumption of industrial enzymes for the food industry amounted to approximately US\$300 × 10^6 in 1990, the main industries being starch and dairy. Brewing, wine and juice and baking together used less than the dairy industry even though significant potential exists for increasing usages of industrial enzymes in these industries.

The total growth (including non-food applications) in volume of the enzyme business from 1974–1986 was 10–15% per year although no major new applications emerged in this period. From 1980–1990 the growth can be estimated as 5–10% per year. Lately new non-food applications have been developed within the textile business. The major worldwide sales value of enzyme product types used by the food industry can be seen in Table 2.1.

2.2 Enzymes for starch modification

Depending on the enzymes used and the reaction conditions employed various valuable products may be produced to suit nearly any particular requirement of the food industry. Syrups and modified starches with different compositions and physical properties can be obtained. The syrups

Table 2.1 Estimated world sales value of food enzymes by product type, 1990

Enzyme	Consumption (10^6 US\$)
Rennet (animal and microbial)	75
Glucose isomerase	40
Glucoamylase	75
α-Amylase	50
Papain	8
Trypsin	8
Other food proteases	8
Invertase	8
Pectinase	7
Other (β-glucanase, cellulase, dextranase, glucose oxidase, hemicellulase, lactase, lipase, pullulanase)	20

are used in a wide variety of foodstuffs: soft drinks, confectionery, meat products, baked products, ice cream, sauces, baby food, canned fruit, preserves, etc.

The major steps in conversion of starch are liquefaction, saccharification and isomerization. In simple terms, the further a starch processor proceeds, the sweeter the syrup that can be obtained.

2.2.1 *Tailor-made glucose syrups*

Glucose syrups, in commercial practice called glucose, are obtained by hydrolysing starch (mainly from wheat, maize (corn) and potato), by a process which cleaves the bonds linking dextrose units of the starch chain. The method and the extent of hydrolysis (conversion) affects the final carbohydrate composition, and by that, many functional properties. The degree of hydrolysis is commonly defined as the dextrose equivalent (DE), expressing the reducing power as a percentage of pure dextrose, calculated to dry weight basis.

Originally, acid conversion was used to produce glucose syrups. Today, because of their specificity, enzymes are frequently used to predetermine exactly how the hydrolysis will take place. In this way, tailor-made glucose syrups with specific carbohydrate spectra can be manufactured.

The carbohydrate spectra can be analysed by different techniques, such as high performance liquid chromatography (HPLC) and gel permeation chromatography (GPC). They give information about the molecular weight distribution and overall carbohydrate composition of the glucose syrups, which, at the same time, defines and characterizes the product type (e.g. high maltose syrup). Although these techniques help to optimize glucose syrup profiles for specific applications, indirect methods, such as viscosity measurements, are also important means of reaching the goal of a tailor-made product.

A slurry of the starch is cooked in the presence of a heat-stable bacterial endo-α-amylase. The enzyme hydrolyses the α-1,4-glycosidic bonds in pregelatinized starch, whereby the viscosity of the gel rapidly decreases and the so-called maltodextrins are produced. The process may be terminated at this point, the solution purified and dried, and the maltodextrins utilized as bland tasting functional ingredients in, for instance, dry soup mixes, infant foods, sauces, gravy mixes and so on. Further hydrolysis by means of amyloglucosidase leads to the formation of sweet tasting, fermentable sugars. Sweet starch hydrolysates with special functional properties may be obtained by using fungal α-amylase either alone or in combination with amyloglucosidase. Alternatively a vegetable β-amylase may be used to increase the yield of maltose.

Many non-food products also produced by fermentation rely on the use of enzymically modified starch products. For instance, enzyme-hydrolysed starches are used for the production of alcohol, polyols, ascorbic acid, enzymes, lysine and penicillin.

2.2.2 *Regulatory considerations*

National authorities have legislated for and regulated the use of enzymes in the food industry. The joint FAO/WHO Expert Committee on Food Additives (JECFA) and the Food Chemicals Codex (FCC) have published guidelines for the use of enzymes as food additives. The Association of Microbial Food Enzyme Producers (AMFEP) in Europe and the Enzyme Technical Association (ETA) in the United States, work nationally as well as internationally for the harmonization of regulations. In the European Community, common guidelines are being drafted for the evaluation of enzymes for food and feed uses.

Obviously, a microbial source of a food enzyme must be non-pathogenic and non-toxic. Manufacturers of microbial food enzymes have always selected their production microorganisms from the safe end of the spectrum of available sources. Consequently, a few species have acquired a record of safe use as sources of a wide variety of food enzymes. For enzymes from recombinant microorganisms, the primary regulatory status is determined by the host organism, the donor organism, and any vectors involved in the genetic transfer.

The majority of food enzymes are used as processing aids, and have no function in the final food. In this case they do not need to be declared on the label, as they will not be present in the final food in any significant amount. Some enzymes (a few) are used both as processing aids and as food additives. When used as additives, they must be declared on the food label using the appropriate class name, e.g. preservative or antioxidant, E-number and generic name like lysozyme or glucose oxidase.

2.2.3 *Quality assurance of industrial enzymes*

AMFEP has defined good manufacturing practice (GMP) for microbial food enzymes. The most important element is to ensure pure culture of the production organism. Product specifications for microbial food enzymes have been established by JECFA and FCC. They prescribe the absence of contaminants like arsenic, heavy metals, lead, coliforms, e.g. *E. coli* and *Salmonella*. Furthermore they prescribe the absence of antibacterial activity and mycotoxins for fungal enzymes only.

Commercial enzyme products are usually formulated in aqueous solutions or processed to non-dusty dry products (granulated or micro-granulated). Both types of preparations must be formulated to suit the final application. Requirements mainly regarding the storage stability (e.g.

enzyme activity stability, microbial stability, physical stability and formulations) as significant quality parameters of industrial enzyme products must be taken into consideration by both the producer and the user.

2.3 Processing and enzymology

The corn wet-milling industry is utilizing modern enzyme technology to a very large extent. Current developments focus on refinements of the basic enzyme conversion processes in order to improve process yields and efficiency. A survey of the major steps in conversion of starch is given in Figure 2.1. The particular steps are described below.

2.3.1 Starch liquefaction

As native starch is only slowly degradable by use of α-amylases, gelatinization and liquefaction of a 30–40% dry matter suspension is needed to make the starch susceptible to enzymatic breakdown. The gelatinization temperature is different for different starches (Lii and Chang, 1991). Maize starch is the most widespread source followed by wheat, tapioca and potato. A good liquefaction system is necessary for successful saccharification and isomerization. For liquefaction an efficient temperature stable α-amylase is added to the starch milk. The mechanical

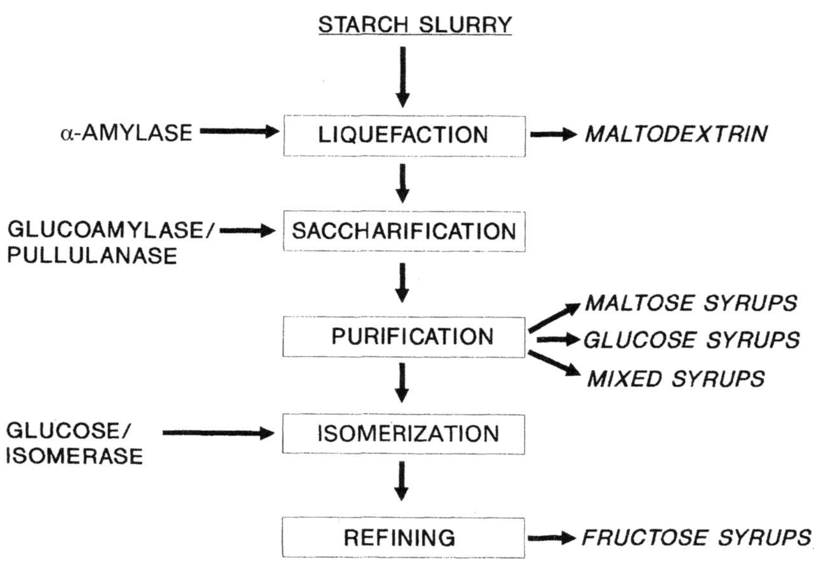

Figure 2.1 Major steps in enzymic starch conversion.

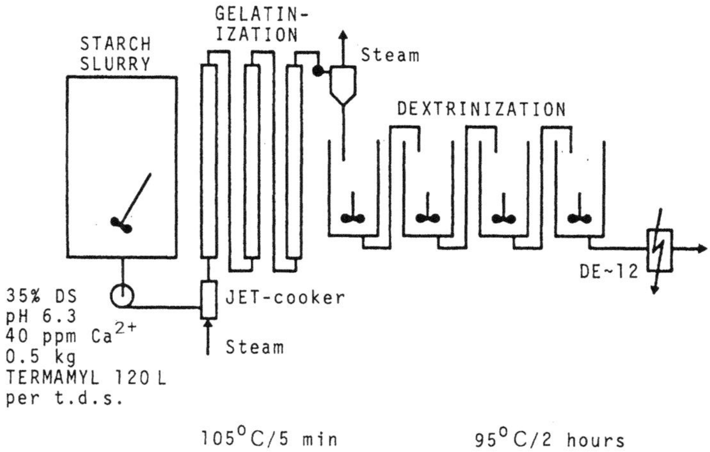

Figure 2.2 Starch liquefaction process. Termamyl is a bacterial thermophillic α-amylase produced by Novo Nordisk (Application sheet, 1990).

method of performing the liquefaction may be by use of stirred tank reactors, continuous stirred tank reactors (CSTR) or a jet cooker.

In most plants the starch liquefaction is made in a single dose jet-cooking process as shown in Figure 2.2. The α-amylase Termamyl® is metered into the starch slurry after pH adjustment, and the slurry is pumped through a jet cooker. Live steam is injected here to raise the temperature to 105°C, and subsequent passage through a series of holding tubes provides a 5 min residence time, which is sufficient to fully gelatinize the starch. By flashing the partially liquefied starch the temperature is reduced to 90–100°C and the enzyme is allowed to react further at this temperature for 1–2 h until the required DE is obtained.

The jet-cooking process is normally carried out under the following conditions. Dry substance (DS) 30–35% (w/w), pH 6.0–6.5 and a Ca^{2+} content of 20–80 ppm (w/v). The dose is typically between 0.5 and 0.6 kg Termamyl 120L per ton of starch (DS). The stability depends on temperature, Ca^{2+} concentration, ion content and DE. Rosendal *et al.* (1979) described stability data from pilot plant trials, from where it was possible to estimate optimum design parameters for the use of Termamyl in individual processes, thus minimizing the dosage of enzymes. The relationship between half life ($T_{½}$) and temperature was derived from an Arrhenius-type equation (plot of half life versus the reciprocal absolute temperature). It was assumed that the stability under other conditions could be expressed as the standard half life multiplied by correction factors for each parameter:

$$T_{½} = T_{½}^{*} \times K_1\,(\text{pH}) \times K_2\,(\text{DS}) \times K_3\,(\text{Ca}) \times K_4\,(\mu\Omega)$$

Table 2.2 Stabilities of Termamyl (in half-lives)

Reaction parameters		Half life (min)
Temperature (°C)	110	15
	105	50
	100	200
	95	700
Dry substance (% w/w)	35	50
	30	30
	25	20
pH	6.5	200
	6.0	110
	5.5	40
Calcium concentration (ppm Ca^{2+})	70	200
	40	110
	10	40

From a practical point of view stabilities of Termamyl (expressed in half lives) are a function of temperature, dry substance (DS), pH and calcium concentration (Table 2.2).

From Table 2.2 it is possible to deduce the enzyme stability for nearly any set of conditions. The values for the stabilities shown under 'temperature' are simply corrected for changes in DS, pH and Ca^{2+} in the proportions found under the respective headings. For example, the stability of Termamyl at 105°C, 30% DS, pH = 6.0 and 10 ppm Ca^{2+} can be calculated to be:

$$50 \times 30/50 \times 110/200 \times 15/50 \simeq 5 \text{ min half life}$$

Addition of Ca^{2+} up to a total of 40 ppm will increase the half life to 16 min.

The enzyme activity decay follows a first order reaction scheme according to the formula:

$$A_t = A_0 \times \exp - \frac{\ln 2}{T_{1/2}} \times t$$

where A_t = the residual activity at time t, A_0 = the activity at time 0, $T_{1/2}$ = the enzyme stability (half life) (min), t = the reaction time (min).

Hence, the residual activity after 5 min of reaction at the conditions used in the above example (half life = 16 min):

$$A_5 = 100 \times \exp\left(-\frac{\ln 2}{16} \times 5\right) = 80\%$$

In general liquefaction is carried out at pH 6.0–6.5. Lower pH values may be used in such cases, where improved stability can be obtained by other means, e.g. by a reduced reaction temperature. Higher pH values should be avoided due to increased byproduct formation (e.g. maltulose precursors). Maltulose formation is enhanced at high temperatures, high pH and long residence times, and is more pronounced at increasing DEs. One advantage of Termamyl is that the temperature is increased only to 105°C at a very low DE compared with hydrochloric acid (at pH 1.8–2), where a temperature of 140°C is used to obtain DE 7–8. A *B. subtilis* α-amylase treatment at 85°C is sometimes used for thinning of the starch before pressure cooking with acid.

The above mentioned process (Figure 2.2) can be used for the production of maltodextrins (DE 15–20). In this case, a higher enzyme dose and/or a longer residence time is needed.

Without too much success, cooking extruders have also been studied for the liquefaction of starch but due to the high temperature in the extruder, inactivation of the enzyme demands 5–10 times higher dosage levels than in a jet cooker (Chouvel *et al.*, 1983).

2.3.2 α-Amylases

During liquefaction the α-1,4-linkages are hydrolysed at random. This reduces the viscosity of the gelatinized starch and increases the dextrose equivalent (DE), a measure of the degree of the hydrolysis of the starch. The liquefaction is carried out in such a way as to give the required DE for the subsequent process. For saccharification to dextrose, a DE of 8–12 is commonly used. Higher DE values are often necessary for maltodextrin production. The maximum DE obtainable is about 40.

Industrially important α-amylases are those made from *Bacillus licheniformis*, *Bacillus subtilis* and *Aspergillus oryzae*. In Table 2.3 some application conditions and a few comparative characteristics are shown for various enzymes available for starch processing. The four amylases from Novo Nordisk are described here.

2.3.2.1 Bacterial thermophilic α-amylase. Termamyl is an *endo*-amylase which hydrolyses the α-1,4-linkages in starch (amylose and amylopectin) almost at random. The breakdown products formed are mainly soluble dextrins and oligosaccharides. In a concentrated solution of starch, the hydrolysis results in a rapid viscosity reduction. In consequence Termamyl is often referred to as a 'liquefying amylase'. Termamyl is available in liquid form with a standardized activity of 60 or 120 KNU g^{-1}. One KNU (*K*ilo *N*ovo α-amylase *U*nit) is the amount of enzyme that hydrolyses 5.26 g of (standard) starch per h at 37°C, pH = 5.6 under standard conditions.

Table 2.3 Industrially important starch degrading enzymes

Enzyme	Microorganism	Application temperature range (°C)	Application pH range	Minimum Ca^{+} dosage (ppm)
Bacterial mesophilic α-amylase	*Bacillus subtilis*	80–85	6–7	150
Bacterial thermophilic α-amylase	*Bacillus licheniformis*	95–105	6–7	20
Fungal α-amylase	*Aspergillus oryzae*	55–70	4–5	50
Amyloglucosidase	*Aspergillus niger*	55–65	3.5–5	0
Pullulanase	*Bacillus acidopollulyticus*	55–65	3.5–5	0

2.3.2.2 Fungal α-amylase. Fungamyl® is a fungal *exo*-amylase, which hydrolyses the α-1,4-linkages in liquefied starch (amylose and amylopectin); a prolonged reaction results in the formation of large amounts of maltose. Fungamyl is used for production of high maltose syrups or high conversion syrups (see section 2.3.5.2). Fungamyl is available in liquid form with a standardized activity of 800 FAU g^{-1}. One FAU (*F*ungal *a*lpha-amylase *u*nit) is the amount of enzyme that hydrolyses 5.36 g of (standard) starch per h at 37°C, pH = 4.7 under standard conditions.

2.3.2.3 Amyloglucosidase. AMG® is a glucoamylase (*exo*-amylase), which hydrolyses α-1,4-linkages as well as α-1,6-linkages in liquefied starch (amylose and amylopectin). The hydrolysis proceeds in a stepwise manner. The breakdown product formed is glucose, which has been split off from the non-reducing end of the substrate molecule. Maltotriose and in particular maltose are hydrolysed at a lower rate than higher saccharides, and 1,6-linkages are broken down more slowly than 1,4-linkages. Eventually, a practically complete conversion of starch into glucose is obtained. In consequence AMG is referred to as a 'saccharifying amylase'. AMG is available in liquid form with a standardized activity of 200, 300 or 400 AGU g^{-1}. One AGU (*A*myloglucosidase *U*nit) is the amount of enzyme that hydrolyses 1 μmol of maltose per min at 25°C under standard conditions.

2.3.2.4 Pullulanase. Promozyme® is a pullulanase, which hydrolyses the α-1,6-glycosidic bonds in starch which has been partly hydrolysed by α-amylase, provided that there are at least two glucose units in the side chain. Promozyme is therefore well suited for debranching starch after liquefaction. Promozyme is available in liquid form with a standardized activity of 200 PUN g^{-1}. One PUN (*P*ullulanase *u*nit *N*ovo) is the amount of enzyme that hydrolyses pullulan liberating reducing carbohydrate with a reducing power equivalent to 1 μmol glucose per min, using a substrate concentration of 0.2% pullulan at 40°C, pH = 5.0 under standard conditions.

2.3.3 *β-Amylases*

β-Amylases are *exo*-enzymes, which attack amylose chains resulting in efficient successive removal of maltose units from the non-reducing end. In the case of amylopectin the cleavage stops two to three glucose units from the α-1,6-branching points. *β*-amylase is used for the production of maltose syrups and in breweries for adjunct processing. The most important commercial products are made from barley or soybeans. Spezyme®BBA 1500, from Genencor International is a barley α-1,4-D-glucan maltohydrolase (E.C. 3.2.1.2) and is commonly referred to as *β*-amylase. It is extracted from barley grain. *β*-Amylase is an *exo*-amylase that catalyses the release of successive maltose units by hydrolysing α-1,4-D glycosidic linkages from the non-reducing end of the dextrin chain. The enzyme is free of α-amylase activity, therefore virtually no glucose is formed.

A purified maltogenic amylase (Maltogenase™) is produced by Novo Nordisk from a genetically modified strain of *Bacillus subtilis* (host), which has received the gene for maltogenic amylase from a strain of *Bacillus stearothermophillus* (donor).

2.3.3.1 Maltogenase™. Maltogenase is a maltogenic amylase (*exo*), which hydrolyses α-1,4-glycosidic linkages from the non-reducing end of the maltodextrin chain. Maltodextrins, oligosaccharides and maltotriose are hydrolysed mainly to maltose. Maltogenase is available in liquid form with a standardized activity of 4000 MANU g^{-1}. One MANU (*M*altogenic *A*mylase *N*ovo *U*nit) is the amount of enzyme which under standard conditions hydrolyses one micromole of maltotriose per min at 37°C, pH 5.0. The substrate concentration under standard conditions is 10 g maltotriose/l.

2.3.4 *Isoamylase and pullulanase*

Isoamylase (glycogen-6-glucanohydrolase) and pullulanase (pullulan-6-glucanohydrolase) hydrolyse α-1,6-glycosidic bonds of starch (section 2.3.2). When amylopectin is treated with pullulanase, linear amylose fragments are obtained. Using heat- and acid-stable pullulanase in combination with saccharification enzymes makes the starch conversion reactions more efficient (Norman, 1982).

2.3.5 *Saccharification of liquefied starch*

Maltodextrin (DE 15–25) produced from the liquefied starch is commercially valuable for its rheological properties. Maltodextrins are used in the food industry as fillers, stabilizers, thickeners, pastes and glues. When saccharified by further hydrolysis using amyloglucosidase or fungal α-amylase a variety of sweeteners can be produced, having dextrose equivalents in the ranges 40–45 (maltose), 50–55 (high maltose) and 55–70 (high conversion syrup) (Reichelt, 1983). Applying a series of enzymes including *β*-amylase, glucoamylase and pullulanase as debranching enzymes, intermediate-level conversion syrups having maltose contents close to 80% can be produced (Norman, 1982).

A high content of glucose (95–97%) may be produced from most starch raw materials (corn, wheat, potatoes, tapioca, barley and rice). The action of amylases and debranching enzymes is shown in Figure 2.3.

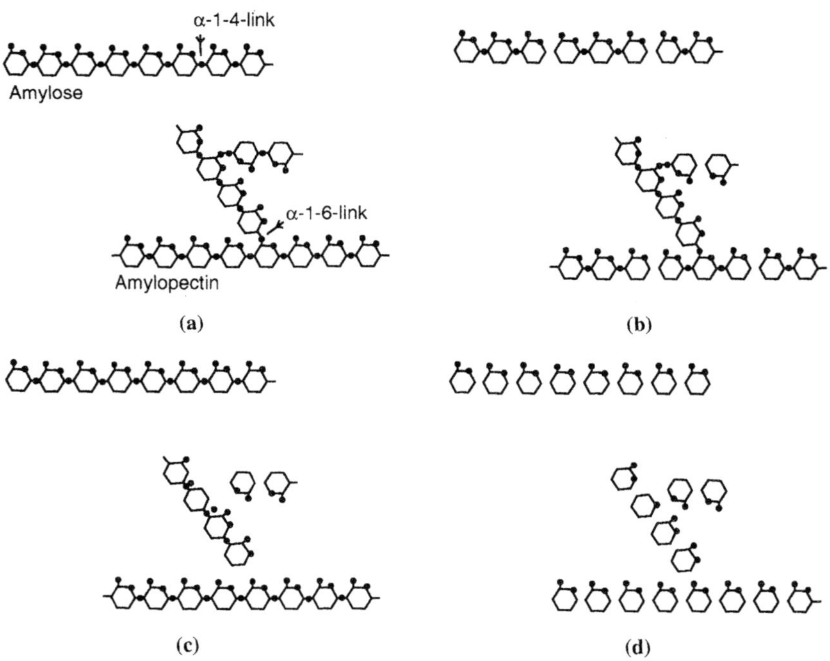

Figure 2.3 Effect of the action of starch degrading enzymes (shown schematically). (a) Amylose and amylopectin structures. (b) Effect of fungal α-amylase on amylose and amylopectin. (c) Effect of a debranching enzyme on amylose and amylopectin. (d) Effect of amyloglucosidase and debranching enzyme on amylose and amylopectin.

2.3.5.1 Use of maltose syrups. Maltose syrups are characterized by having a low content of glucose and a high content of maltose. Because of the low content of glucose, high maltose syrups show a low tendency to crystallize and they are relatively non-hygroscopic. Various specifications (maltose and glucose content) of saccharification products and some applications of maltose syrups are shown in Table 2.4

2.3.5.2 Production of maltose syrups. Maltose syrups are produced by saccharifying liquefied starch with maltogenic *exo*-enzymes. Enzymes for this purpose are fungal α-amylase (e.g. Fungamyl, Maltogenase, Promozyme) or barley *β*-amylase (e.g. SpezymeBBA). Saccharification is conducted at 55–65°C, pH 4.8–5.5, and at a solids level DS 30–40%. The sugar spectrum obtained depends on the enzyme dosage and on the reaction time. The enzyme dosage may be in the range of 0.3–3 kg enzyme products per ton of starch (DS) and the reaction time may be in the range of 20–42 h. In the following examples saccharification systems and reaction conditions are shown for the various maltose syrups mentioned in Table 2.4. The influence of reaction parameters on the composition of the syrups is illustrated by graphs, bar diagrams and carbohydrate spectra. All these data are from the enzyme development laboratories and pilot plant at Novo Nordisk A/S.

2.3.5.2.1 High maltose syrup—by Fungamyl®. High maltose syrups are produced by saccharifying enzyme-liquefied starch (DE 10–20 may be used) with Fungamyl 800L (section 2.3.2). The process should be carried out at 55°C and pH 5.5. It is important that the pH does not drop below pH 5, as the stability of Fungamyl is poor at lower pH values. Starch, dextrins and oligosaccharides have a stabilizing effect on the enzyme.

Saccharification with Fungamyl is normally carried out at 30–40% DS, and no adjustment of the dry substance concentration after liquefaction is necessary. Calcium ions also stabilize Fungamyl. A satisfactory stability is achieved at about 50 ppm Ca^{2+}. Figure 2.4 shows the effect of two dosages (0.2 kg/t DS and 0.4 kg/t DS) of Fungamyl 800L and Figure 2.5 shows the effect of the temperature. The highly significant effect of the two reaction parameters on the composition of the syrup demonstrates the importance of being able to analyse and control the reaction while it is going on.

Figure 2.6 shows one example of the composition of a saccharification carried out by use of 400 g Fungamyl 800L/t DS. Figures 2.7(a) and (b) show the carbohydrate spectra of starch hydrolysed by Fungamyl for 6 and 24 h, respectively. Initially, large amounts of maltotriose are formed, but as the hydrolysis proceeds the level of maltose increases. The normal saccharification time is 24–48 h. It is usually not necessary to inactivate the enzyme after saccharification as the reaction is self terminating.

Table 2.4. Various specifications of saccharification products and some applications

Saccharification product	Maltose content (%)	Glucose content (%)	Typical application: Industry	Typical application: Products and effects
High maltose syrup – by Fungamyl®	50–55	2–5	Confectionery	Moisture and texture control in soft confectionery
High maltose syrup – produced at 65°C	50–55	8–10	Baking	Moisture retention and colour control in final products
			Frozen desserts	Control of softness and freezing characteristics
High maltose syrup specially produced at 60 to 65°C	55–65	8–12	Brewing	Control of fermentation via balanced fermentable sugar spectrum High percentage of fermentable sugars (97+)
Very high maltose syrup	70–75	0–3	Ice cream	Control of softness and freezing characteristics
Ultra high maltose syrup	82–88	5–9	Candy fruit	No crystallization Less hygroscopic product
High conversion syrup[a]	30–37	35–43	Jam and jellies	Viscosity and osmotic profiles Sweetness control
			Brewing	Fermentation control
			Canning	Bodying and sweetness in fruit canning
			Confectionery	Candies for high sucrose replacement Less hygroscopic products
			Baking	Moisture retention and colour control in final products
			Soft drinks	Stabilization of flavour profile during shelf life
Glucose syrup[b]	1–2	94–97	Baby food, dietetic foods	Instant energy source
			Baking	Dough properties and caramelization

[a] High conversion syrups should be stable enough to resist crystallization at temperatures above 4°C and 80–83% DS. The content of glucose in these syrups should therefore not exceed 43% on DS basis. [b] Apart from the food industry, syrups and crystalline glucose (dextrose) also have applications in the pharmaceutical and chemical industry.

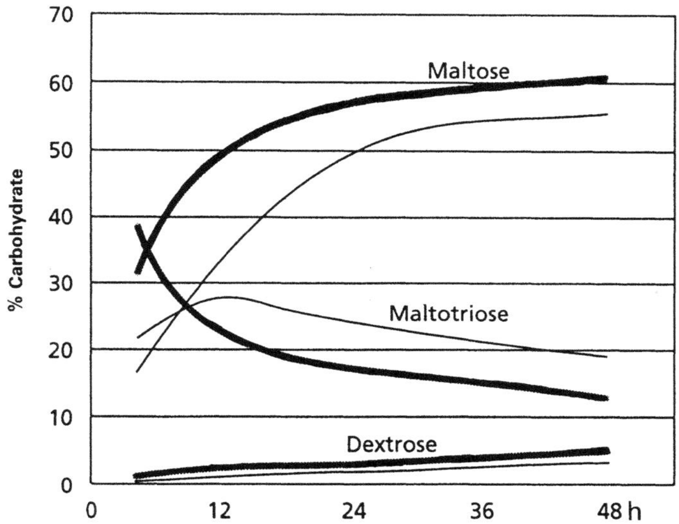

Figure 2.4 Effect of dosage of Fungamyl 800L on the carbohydrate composition of high maltose syrup. ——, 0.2 kg t^{-1} DC Fumgamyl; —, 0.4 kg t^{-1} DS Fungamyl.

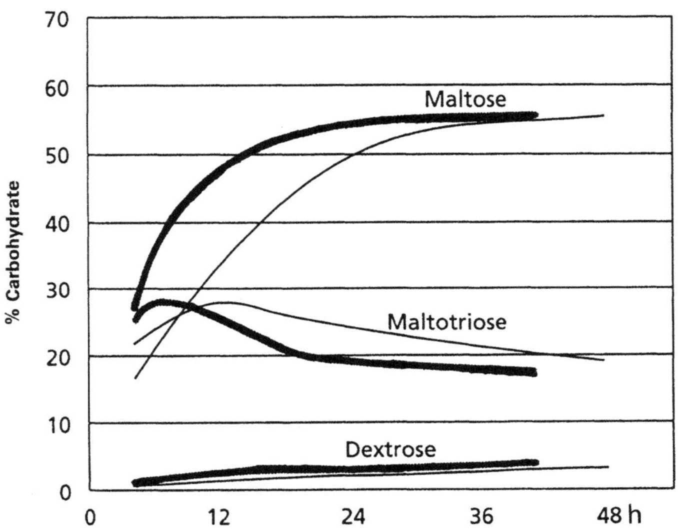

Figure 2.5 Effect of temperature on the carbohydrate composition of high maltose syrup. ——, 55°C, 0.2 kg t^{-1} DS Fungamyl; —, 60°C, 0.5 kg t^{-1} DS Fungamyl.

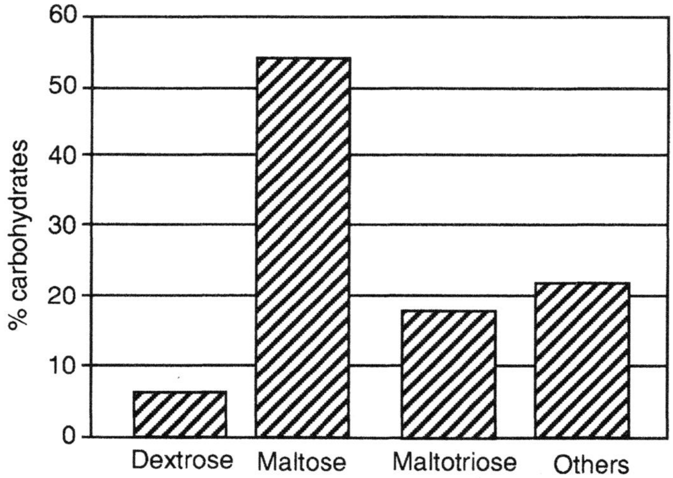

Figure 2.6 Carbohydrate composition of a high maltose syrup. Saccharification time is 22 h using Fungamyl at 55°C.

After saccharification the syrup is filtered and refined by means of activated carbon and, if desired, demineralized by ion exchange and finally concentrated by evaporation.

2.3.5.2.2 High maltose syrup produced at 65°C. A way to increase the maltose content and at the same time reduce the content of maltotriose is to use an unconventional combination of enzymes containing 0.4 kg Termamyl 120L/t DS and Maltogenase and Promozyme in varying doses. The process should be carried out at 65°C and pH 5.0. The carbohydrate spectrum varies acording to the reaction time and the level of addition of enzymes as shown in Table 2.5 and Figure 2.8.

2.3.5.2.3 High maltose syrup specially produced at 60 to 65°C. These brewing syrups can be characterized by 55–65% maltose content and a low content of maltotriose and some glucose. The enzyme-liquefied starch (DE 10–20) can be hydrolysed at 60°C, pH 6 using Maltogenase (0.25–1.0 l/t DS) + Fungamyl (0.4–1.0 kg/t DS), or at 65°C, pH 5 using Maltogenase (0.5–1.0 l/t DS) + Promozyme (0.5–1.0 l/t DS) + 0.25 kg/t DS BAN 240L to all. For BAN, see section 2.3.2. Carbohydrate compositions are shown in Figure 2.9.

2.3.5.2.4 Very high maltose syrup. An extremely low glucose content can be secured by use of malt extract, approx. 400° Lintner (3–4 kg/t DS)

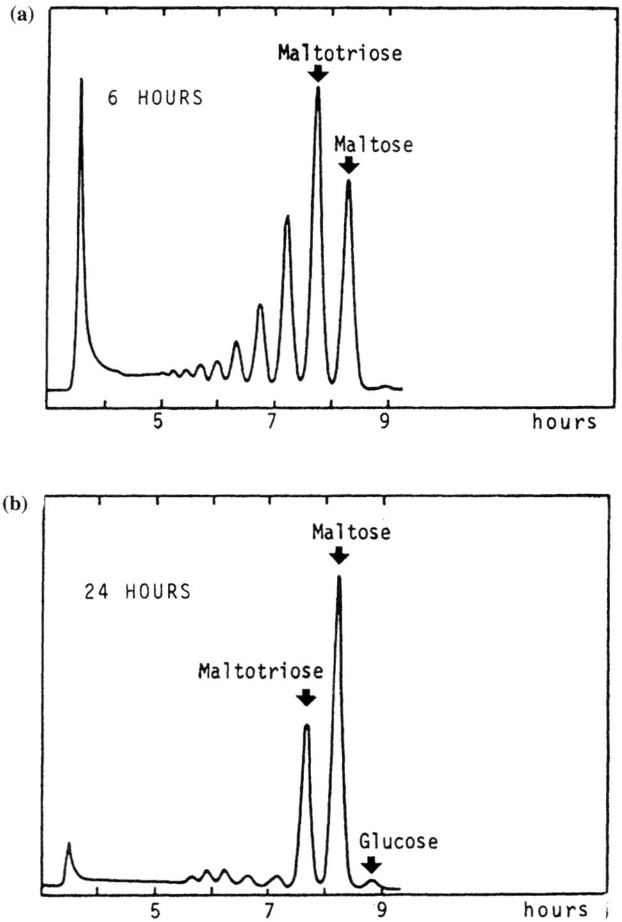

Figure 2.7 Carbohydrate spectra of starch hydrolysates using Fungamyl 800L. (a) 6 h hydrolysis time; (b) 24 h hydrolysis time.

Table 2.5 High maltose syrup produced at 65°C

α-Amylase (Termamyl)+ . . . (l/t DS)	% Glucose		% Maltose		% Maltotriose		Others	
	24 h	42 h	24 h	42 h	24 h	42 h	24 h	42 h
0.4 l Maltogenase + 0.3 l Promozyme	9	11	51	58	12	9	28	22
0.5 l Maltogenase + 0.5 l Promozyme	8	10	54	63	10	8	28	19
1.0 l Maltogenase + 1.0 l Promozyme	9	11	67	75	6	3	18	11

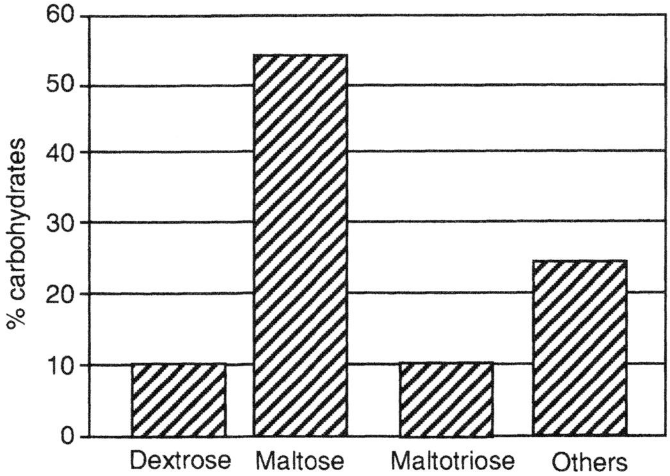

Figure 2.8 Carbohydrate composition of high maltose syrup produced at 65°C using Maltogenase. More maltose results from higher levels of addition.

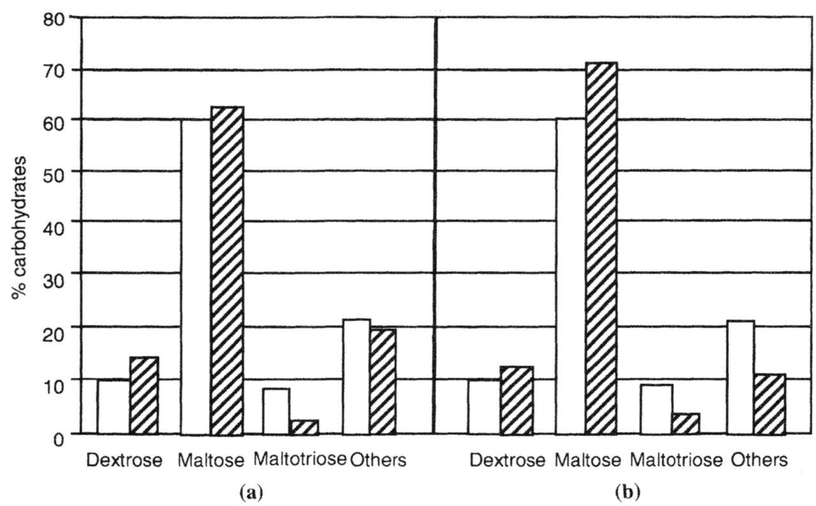

Figure 2.9 High maltose syrup special, 55–60% maltose at 60–65°C. (a) At 60°C by Maltogenase and Fungamyl, (b) at 65°C by Maltogenase, Promozyme and BAN. Sugar composition at □ 60% maltose, ▨ around maximum maltose.

or β-amylase 1500°Lintner (1 kg/t DS) + Promozyme 600L (1 l/t DS). An addition of BAN to the β-amylase saccharification after reaction for one day increases the maltose content. Furthermore, it improves the filterability of the syrup.

The DE after liquefaction influences the final amount of maltose; the

Table 2.6 Very high maltose syrup

Promozyme and . . .	% Glucose	% Maltose	% Maltotriose	% Others
Malt extract	2.5	72	18	7.5
β-Amylase	0.5	72	13	15.5
β-Amylase + BAN[a]	1	75	15	9.0

[a]BAN added after 1 day.

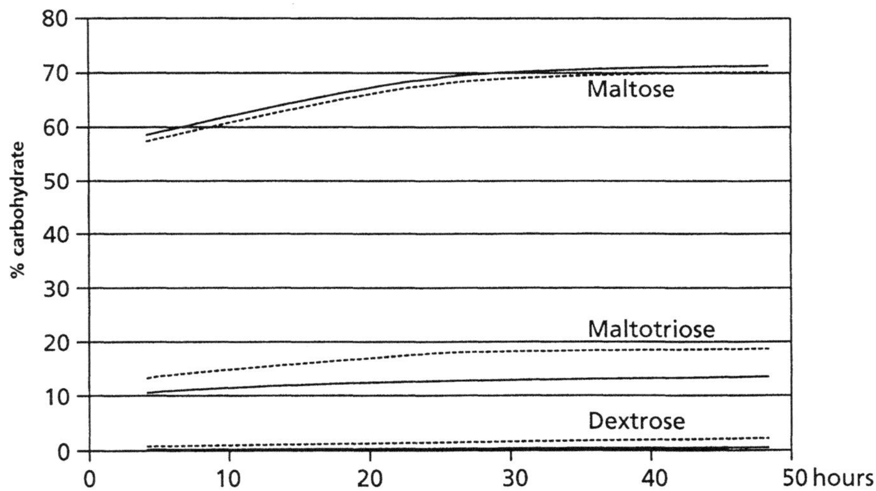

Figure 2.10 Progress of saccharification with time. - - -, Promozyme + malt extract; ——, Promozyme + *β*-amylase.

higher the DE, the lower percentage of maltose. So, in this case certain constraints are set for the reaction parameters in order to obtain the very high maltose level. The enzyme-liquefied starch must have a maximum DE of 10, at DS = 30%. The temperature must be adjusted to 58°C and pH to 5.5. The composition shown in Table 2.6 will thus be obtained after 2 days. The time course of the saccharification is shown in Figure 2.10, and it can be seen that the increase in percentage maltose is minimal after 30 h.

2.3.5.2.5 Ultra high maltose syrup. The procedure for manufacture of ultra high maltose syrups requires, as above, that the enzyme-liquefied starch has a maximum DE of 10, at DS = 30%. The temperature must also be adjusted to 58°C and pH to 5.5. These syrups are characterized by having an ultra low content of maltotriose and, as indicated, an ultra high content of maltose. In fact these syrups have the highest maltose content seen for enzymatically produced sugars. The enzyme systems used are

Table 2.7 Ultra high maltose syrup

Time	% Glucose	% Maltose	% Maltotriose	% Others
after 24 h	6.5	82	6	5.5
after 48 h	7.5	83	4	4.5
after 72 h	8.5	84	2.5	5.0

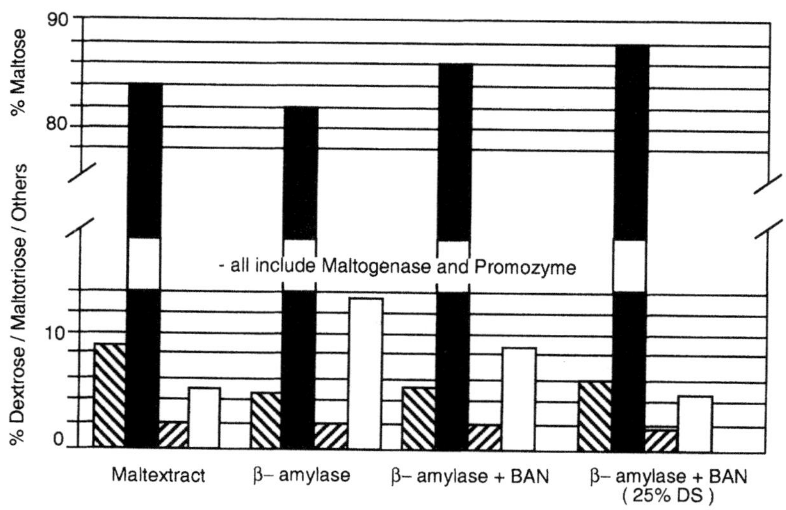

Figure 2.11 Influence of various enzyme combinations for production of ultra high maltose syrup. ▧, Dextrose; ■, maltose; ▨, maltotriose; □, other.

Maltogenase (1.5 l/t DS) + Promozyme (1.0 l/t DS) + malt extract (1 kg/t DS). If malt extract is replaced by *β*-amylase, the maltose content will be around 2% lower. This can, however, be more than counteracted by the addition of BAN, so that the final maltose content will be 1 to 2% higher than the figures shown in Table 2.7. In Figure 2.11 the influence of various enzyme combinations, using the above mentioned enzyme products, are shown.

2.3.5.2.6 High conversion syrup. High conversion syrups are produced from enzyme-liquefied starch or from acid-liquefied starch by using a combination of Fungamyl 800L and AMG 200L. Higher maltose and lower glucose levels are obtained with enzyme-liquefied starch. The following conditions can be used for the production of high conversion syrups.

The process may be carried out at 55°C and pH 5.0. The dry substance should be in the range 30–40% and no adjustment of dry substance concentration after liquefaction is, therefore, necessary. The saccharification time is normally about 40 h, but shorter or longer reaction times may be used with proportionally higher or lower enzyme doses.

It is very important to inactivate the AMG when the required DE and sugar composition have been reached. If the reaction is allowed to proceed more glucose will be produced, which crystallizes when the syrup is concentrated and stored. The reaction may be stopped by heating the syrup to 80°C for about 40 min or to 85°C for about 5 min. The syrup is then filtered, carbon treated and, if desired, ion exchanged.

The sugar composition of the final syrup may be altered by adjusting the ratio between the Fungamyl and the AMG dosages as shown in Table 2.8 for enzyme-liquefied starch and in Table 2.9 for acid-liquefied starch. As an example the 'standard' high conversion syrup with 41% glucose and 37% maltose can be obtained from enzyme-liquefied starch with 100 g

Table 2.8 Effect of AMG and Fungamyl doses on syrup composition for enzyme-liquefied starch

	AMG ml/t DS											
	0		37.5		75		112.5		150		187.5	
Fungamyl g/t DS	G[a]	M[b]	G	M	G	M	G	M	G	M	G	M
0					37	11	47	15	57	12	65	9
50	1	33	15	35	30	35	41	33	49	30	51	24
100	2	45	11	46	21	44	31	42	41	37	48	31
150	3	51	11	51	20	49	29	45	38	40	45	35
200	3	54	14	53	24	49	34	44	43	37	50	31
250	4	54	14	54	25	51	34	45	43	38	49	33

[a] % glucose (G), [b] % maltose (M). Substrate, 12 DE enzyme-liquefied starch, 35% DS; temperature, 55°C, pH 5.0; reaction time, 42 h.

Table 2.9 Effect of AMG and Fungamyl dosages on syrup composition for acid-liquefied starch

Fungamyl g/t DSD	AMG ml/t DS 0 G[a]	M[b]	37.5 G	M	75 G	M	112.5 G	M	150 G	M	187.5 G	M
0	18	18	29	21	39	24	49	24	57	21	61	17
50	21	31	31	32	39	31	48	29	55	25	61	20
100	21	36	31	36	40	34	49	30	55	26	60	22
150	23	38	32	37	40	35	49	30	56	25	61	20
200	21	39	30	39	38	36	47	32	54	27	60	21
250	23	40	32	39	40	36	49	31	56	25	61	20

[a] % glucose (G), [b] % maltose (M). Substrate, 44 DE acid-liquefied starch, 45% DS; temperature, 55°C, pH 5.0; reaction time, 42 h.

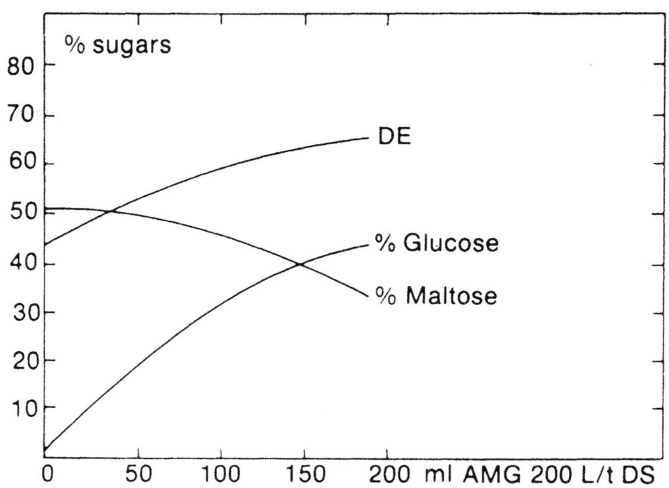

Figure 2.12 Effect of AMG dose on the composition of high conversion syrup. Substrate, 12 DE enzyme-liquefied starch, 35% DS; Fungamyl 800L dose 150 g/t DS; temperature 55°C, pH 5.0; reaction time 42 h.

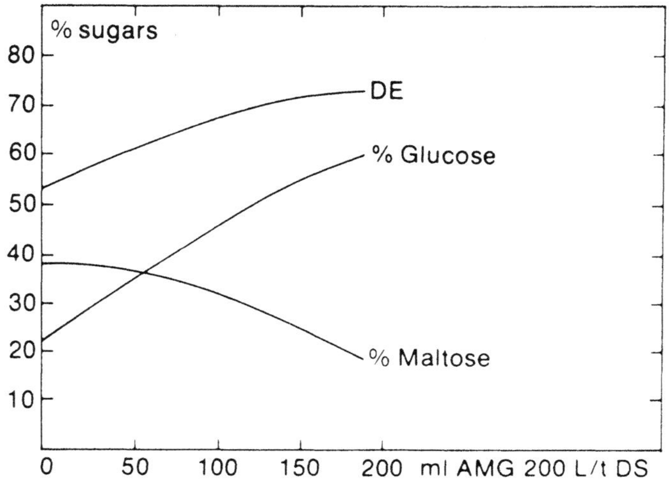

Figure 2.13 Effect of AMG dose on the composition of high conversion syrup. Substrate, 44 DE acid-liquefied starch, 45% DS; Fungamyl 800L dose 150 g/t DS; temperature 55°C, pH 5.0; reaction time 42 h.

Fungamyl 800L and 150 ml AMG 200L/t DS. Further, the tables show that at a constant Fungamyl dosage, DE and glucose levels increase with increasing AMG dosage, while the maltose content tends to decrease. At a constant AMG dosage, DE and glucose contents are almost constant with increasing Fungamyl dosage, while the maltose content increases until it reaches a maximum at a dosage of 150–200 g/t DS.

Figures 2.12 and 2.13 show DE, % glucose and % maltose as functions of AMG dosage at a constant Fungamyl dosage of 150 g/t DS using enzyme or acid-liquefied starch as starting material, respectively.

2.3.6 Production of high dextrose syrups

Amyloglucosidase was first introduced in the early 1960s, and today this saccharification enzyme is considered a commodity enzyme. Amyloglucosidase hydrolyses the α-1,4-linkages rapidly, but during the saccharification process the α-1,6-linkages of the highly branched amylopectin are also hydrolysed but much slower. Normally the dosage of AMG is adjusted to obtain the desired final degree of saccharification, DX, within 48 to 96 h as shown in Figure 2.14. At the start of the saccharification, the rate of dextrose formation is high, but it gradually decreases towards the end of the process. This is partly due to the accumulation of branched dextrins, and partly because the increasing concentration of dextrose accelerates reversion (re-polymerization of dextrose into isomaltose and other saccharides). At a certain point, the rate of reversion outbalances that of

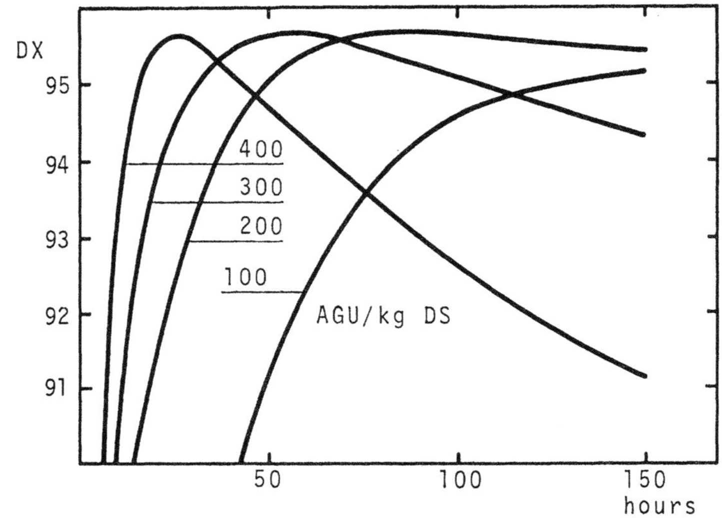

Figure 2.14 Influence of AMG dose on saccharification. Substrate: 12 DE maltodextrin; initial DS 30% w/w; temperature 60°C; pH 4.5.

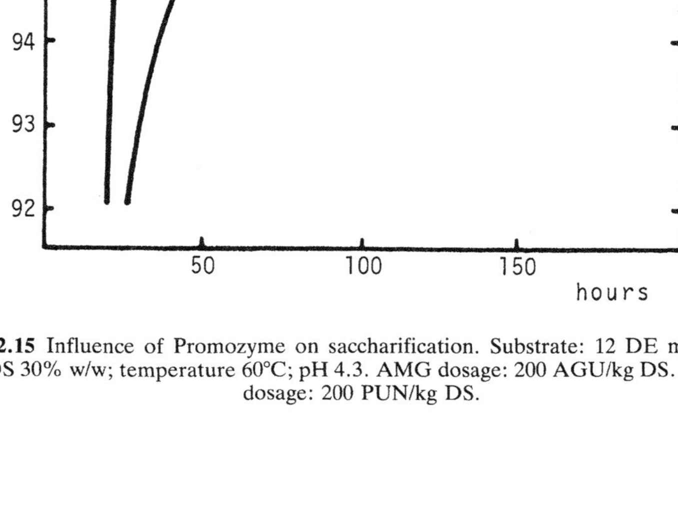

Figure 2.15 Influence of Promozyme on saccharification. Substrate: 12 DE maltodextrin; initial DS 30% w/w; temperature 60°C; pH 4.3. AMG dosage: 200 AGU/kg DS. Promoxyme dosage: 200 PUN/kg DS.

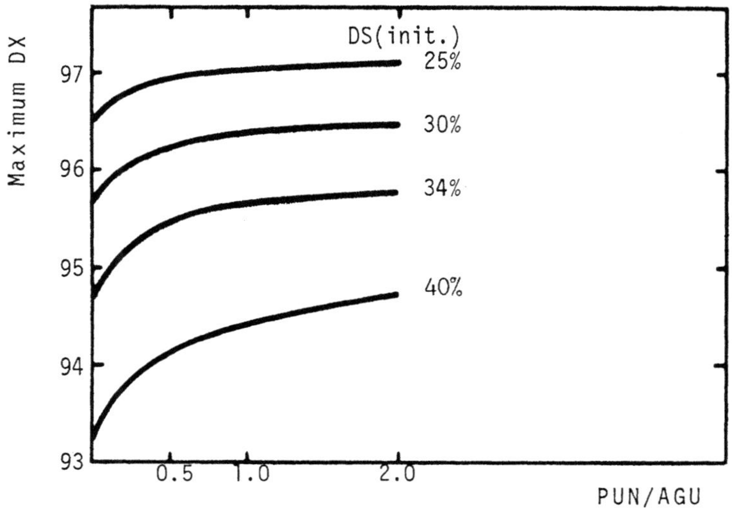

Figure 2.16 Influence of dry substance and PUN:AGU ratio on maximum DX. Substrate: 12 DE maltodextrin; temperature 60°C; pH 4.3. AMG dosage: 100–400 AGU/kg DS. Promoxyme dosage: 0–800 PUN/kg DS.

dextrose formation, and if the saccharification is not stopped at this point (maximum DX), the dextrose level will gradually decrease towards chemical equilibrium (around 85% dextrose at 30% DS, 60°C).

Using a pullulanase like Promozyme, starch is debranched by rapid hydrolysis of the α-1,6-bonds. Using Promozyme together with an amyloglucosidase like AMG at the start of the saccharification the α-1,6-linkages of the branched dextrins are rapidly hydrolysed. In consequence fewer branched oligosaccharides accumulate towards the end of the saccharification. The point at which reversion outbalances dextrose formation is thus shifted towards a higher DX level. This is shown on Figure 2.15, where the effect of Promozyme on saccharification is illustrated.

The maximum DX obtainable with AMG and Promozyme mainly depends on the enzyme dosage ratio (PUN:AGU) and dry substance level, see Figure 2.16. The most dramatic effect of Promozyme was observed at PUN:AGU ratios below 0.5.

During saccharification, the dry substance concentration increases by approximately 10% (relatively). This is because one molecule of water is taken up by the carbohydrate for each molecule of dextrose released by hydrolysis. At high DS levels, the re-polymerization (reversion) is favoured, and consequently the maximum obtainable DX is reduced as shown in Figure 2.16. Normally, an initial DS level of around 30% w/w is chosen as an economic compromise between too low final DX (at higher

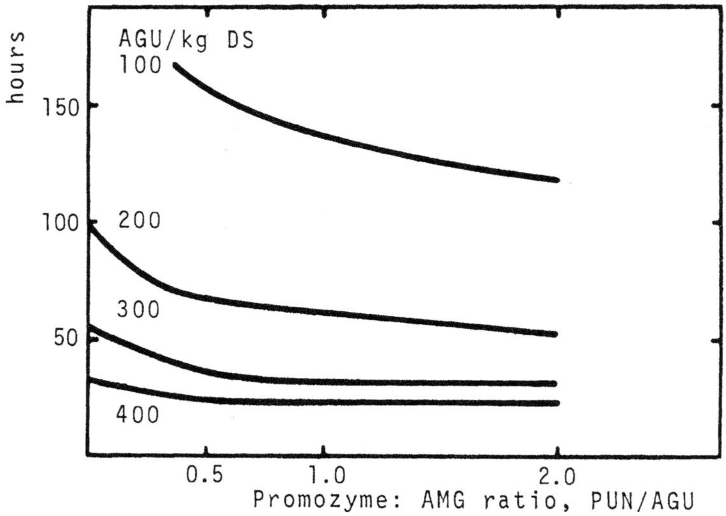

Figure 2.17 Effect of the AMG and Promozyme dosage on saccharification time. Substrate: 12 DE maltodextrin; initial DS: 25–40% w/w; temperature 60°C; pH 4.3. Promozyme dosage: 0–800 PUN/kg DS.

DS) and too high evaporation cost (at lower DS). An increase in DS and thereby a reduction in evaporation cost can be obtained with Promozyme without reducing maximum DX.

From Figure 2.17 it can be seen that the saccharification time necessary to obtain maximum DX depends on the PUN:AGU ratio as well as on the enzyme dosage level. For a given saccharification time, the necessary AGU dosage, and thereby the reversion rate, is reduced when the PUN:AGU ratio is increased.

A saccharification temperature of 60°C is usually applied. At higher temperatures, the stability of the enzyme is reduced. At lower temperatures, the reaction rate will be reduced and the risk of microbial infection will increase.

Optimum enzyme performance is obtained with an initial pH of 4.3 (measured at 60°C). Usually the pH drops slightly during saccharification and the extent of this pH drop should be taken into account when selecting the initial pH set point. The pH should never be allowed to drop below 3.5.

2.3.6.1 Continuous saccharification. It is theoretically possible to obtain exactly the same final DX by continuous saccharification as in a batch process under the same reaction conditions. It would be necessary, however, to use either a plug-flow (tube) reactor or an infinite series of infinitely small continuous-flow stirred tank reactors (CSTR). As illustrated in Figure 2.18, a limited number of saccharification tanks results in a final

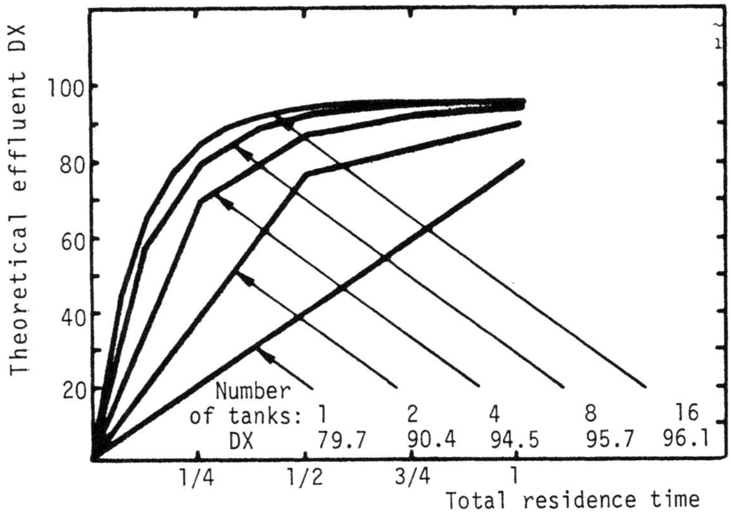

Figure 2.18 Theoretical effluent DX in CSTR series with increasing number of tanks.

DX lower than the theoretical use of an infinite number of infinitely small continuous-flow stirred tank reactors. This is due to retention time variance. A series of minimum eight tanks is recommended.

2.3.6.2 Inactivation and purification. In order to prevent excessive reversion of glucose to isomaltose at the end of saccharification, it is necessary to inactivate the AM within a few hours after maximum DX has been obtained. When Promozyme is used, the AMG dosage and rate of reversion will normally be relatively low, and in consequence the risk of losing dextrose by reversion is reduced.

Ion exchange or carbon treatment of the saccharified liquid will normally remove the AMG/Promozyme activity. Alternatively, the enzyme can be destroyed by heating the liquid to 85°C for about 5 min or to 80°C for about 40 min.

The saccharified liquid is filtered to remove insoluble impurities such as denatured protein and fat (a mixture). In most cases the filtration is followed by carbon treatment for removal of soluble organic impurities and ion exchange.

2.3.6.3 Use of Dextrozyme™ in the production of high dextrose syrup. Dextrozyme® 225/75 L is a balanced mixture of amyloglucosidase and pullulanase. Figure 2.19 shows that when dosed equally Dextrozyme either gives a higher glucose yield (0.3–0.6 DX) than AMG (glucoamylase) or needs a shorter time (10–15%) to reach the same glucose yield as with

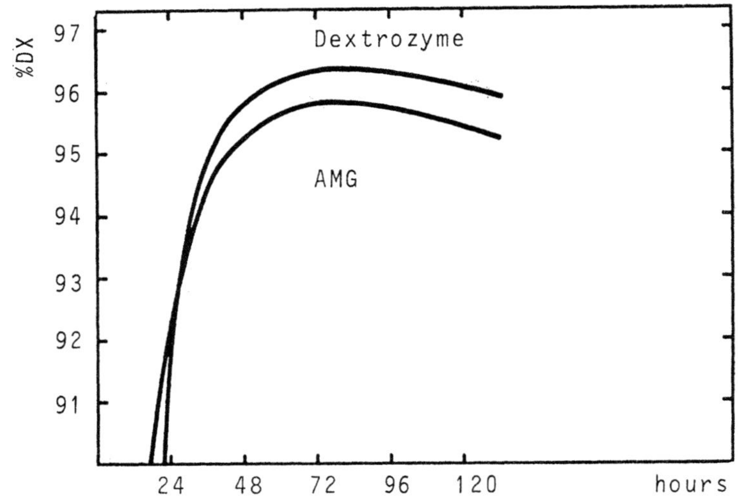

Figure 2.19 Influence of Dextrozyme on saccharification.

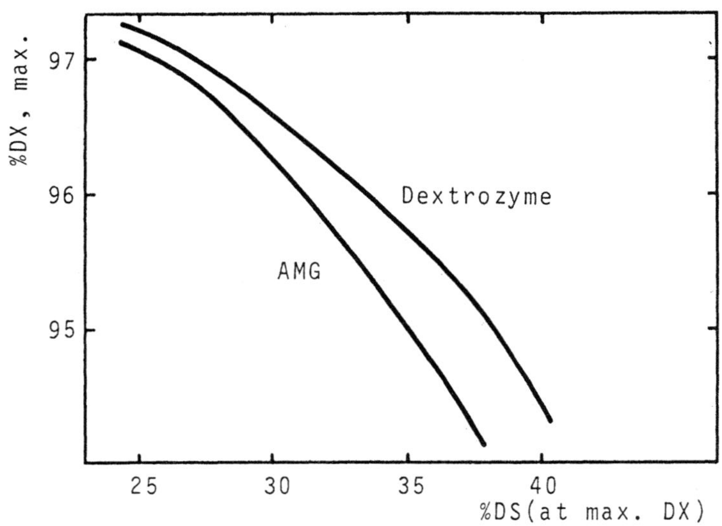

Figure 2.20 Effect of dry substance on maximum obtainable percentage DX.

AMG. This is further illustrated in Figure 2.20, which at the same time illustrates the influence of the dry substance content on the maximum obtainable DX value, the yield of glucose.

The time necessary to reach the maximum obtainable DX depends on the enzyme dosage. The maximum obtainable DX is mainly a function of the dry substance and the nature of the enzyme used. Figure 2.21 illustrates

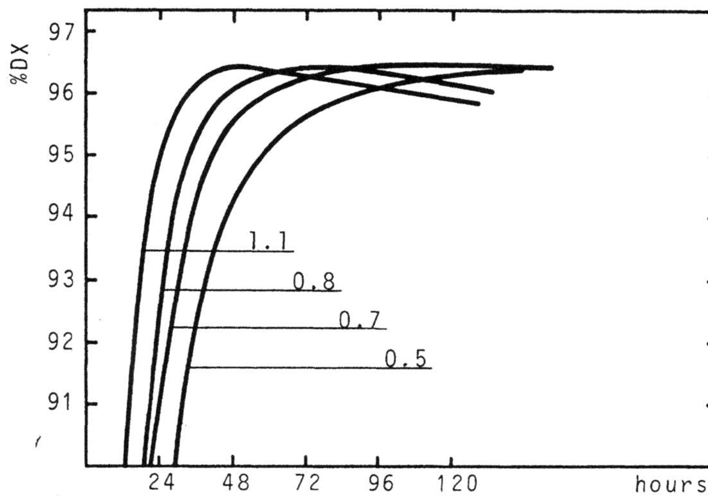

Figure 2.21 Effect of enzyme dosage on saccharification. Litres Dextrozyme 225/75 per ton starch DS.

the influence of enzyme dosage and reaction time on the amount of glucose formed during saccharification. The glucose level starts to fall if the reaction is continued beyond the maximum. This is due to the reverse reaction, which is a condensation of glucose molecules to mainly isomaltose. Also, here it is necessary to stop the reaction, when DX has reached its maximum in order to limit synthesis of isomaltose and to avoid a drop in DX. Purification of the syrup should be performed as in section 2.3.6.2.

The purification processes of saccharified starch may be adapted to each raw material and may be different from plant to plant. When the starch milk is liquefied in the jet cooker (see Figure 2.2) the saccharification process is carried out at 55–65°C, pH 4–4.5 for 24–72 h. The additional steps consist of filtration or centrifugation, ion exchange, isomerization (see later), treatment with activated carbon and evaporation to a storage-stable product.

2.3.6.4 Saccharification in a membrane reactor. A continuous hydrolysis process for the saccharification of liquefied corn starch has been developed using a membrane reactor (Sims and Cheryan, 1992a). A residence time distribution confirmed that the membrane could be modelled as a simple continuous stirred tank reactor (CSTR). Kinetic studies indicated that the continuous reactor operated in the first order region with respect to substrate concentration at substrate concentrations greater than 200 g l^{-1}. The productivities were claimed to be 10 to 20 times higher than those obtained in a batch reactor (Sims and Cheryan, 1992b). All saccharifications were carried out using the amyloglucosidase from *Aspergillus niger* (AMG 200L from Novo Nordisk) at 55°C and pH 4.5 in a dosage of 0.5% v/v (16.6 l t^{-1} of starch). This is a dosage about 15 times higher than normally recommended in batch processes using a saccharification time of 40–70 h. At a residence time of 1 h in the membrane reactor the apparent Michaelis constant (K'_m) was 562 g l^{-1}, which is 2–7 times greater than that obtained in a batch reactor (Sims and Cheryan, 1992b). It seems evident that more work is needed in order to clarify all technical and economical aspects and to test the possibilities of use of membrane technology for saccharification of starch.

2.3.6.5 Enzymes as processing aids in purification of saccharified wheat starch. A minor amount of impurities in the starch milk adheres to the starch granules, especially with wheat starch. To facilitate refining, e.g. filtration operations, cellulases, pentosanases, glucananses, proteases and pectinases are used in some special cases. Wheat starch is known to form precipitates or hazes, which are difficult to filter. Arabinoxylan, pentosans and lysophospholipids are claimed to be responsible for this problem (Konieczny-Janda and Richter, 1991).

2.3.7 Glucose isomerization

Enzymatic isomerization of glucose to fructose was the process that provided a real alternative to white sugar (sucrose) from cane or beet. The commercial product obtained was high fructose corn syrup (HFCS). Two grades of the syrup have established themselves in the world market, HFCS-42 and HFCS-55. They contain 42% or 55% fructose based on dry substance, respectively. These products account for over a third of the carbohydrate sweetener market in the USA. More than 8 million tons of HFCS are produced annually using glucose isomerase. This represents the largest commercial application of immobilized biocatalysts in industry.

2.3.8 The isomerization reaction

Glucose can be reversibly isomerized to fructose. The equilibrium conversion for glucose → fructose is 50% under industrial conditions, and the reaction is slightly endothermic (Tewari and Goldberg, 1984). The reaction is usually carried out at 60°C, pH 7–8. To avoid excessive reaction time, the conversion is normally limited to about 45%.

The isomerization reaction can only be economic by using immobilized enzyme. The reaction parameters in this system have to be as optimal as possible in order to obtain a reasonable yield of fructose. pH must be approximately 7.5 or higher in order to secure high activity and stability of the enzyme. Under these conditions glucose and fructose are rather unstable and decompose easily to organic acid and coloured by-products. To overcome these problems the reaction time must be limited. This is done by using an immobilized isomerase in a fixed-bed reactor process in a column through which glucose flows continuously. The enzyme granules must be rigid enough to prevent compaction during the operation. Sweetzyme T from Novo Nordisk is produced by a mutant of a selected *Streptomyces murinus* strain. The immobilization procedure for Sweetzyme T consists of disruption of a cell concentrate through a homogenizer with a single stage homogenizing valve. The cells are then crosslinked with glutaraldehyde, diluted and flocculated. The concentrated aggregate is extruded and finally fluid-bed dried and sieved (Jørgensen *et al.*, 1988).

2.3.8.1 Isomerization conditions. The activity of the enzyme is determined by the rate of glucose to fructose conversion obtainable at any time. The stability is reflected in the ability of the enzyme to retain its activity during operation. Productivity is the combined effect of activity and stability, and is defined as the weight amount in kg of 42% fructose syrup (dry substance) produced per kg of enzyme throughout its lifetime. The productivity of Sweetzyme T (300 IGIU g^{-1}) normally lies in the range 8000–11 000 kg kg^{-1} depending on feed syrup quality and actual operating conditions.

2.3.8.2 Isomerization temperature. The isomerization temperature strongly affects the enzyme activity and stability as well as the byproduct formation. The standard operating temperature is 60°C. Higher temperatures will result in higher enzyme activity (increased glucose conversion rate), but the stability and productivity will be lower. This means that an increasing activity decay and therefore enzyme consumption will be the result. Furthermore, byproduct formation due to glucose and fructose decomposition will accelerate with increasing temperature.

Lowering the operating temperatures will lead to increased risk of microbial infection as well as to reduced activity (lower glucose conversion rate), but enzyme stability and productivity will be improved. Syrup production plants operating at a high level of hygiene and feed purity may prefer to lower the isomerization temperature towards 55°C or even lower during periods with ample production capacity.

In Figure 2.22, the effect of the isomerization temperature on the activity (syrup flow rate) and productivity of Sweetzyme T is illustrated. Typical performance obtained with Sweetzyme T at 60°C, at 10% residual activity at end-of-run, are:

- Average flow rate: 4–5 kg syrup/kg enzyme per h
- Accumulated productivity: 8000 kg DS/kg enzyme
- The maximal flow rate will be 7–8 kg syrup/kg enzyme per h.

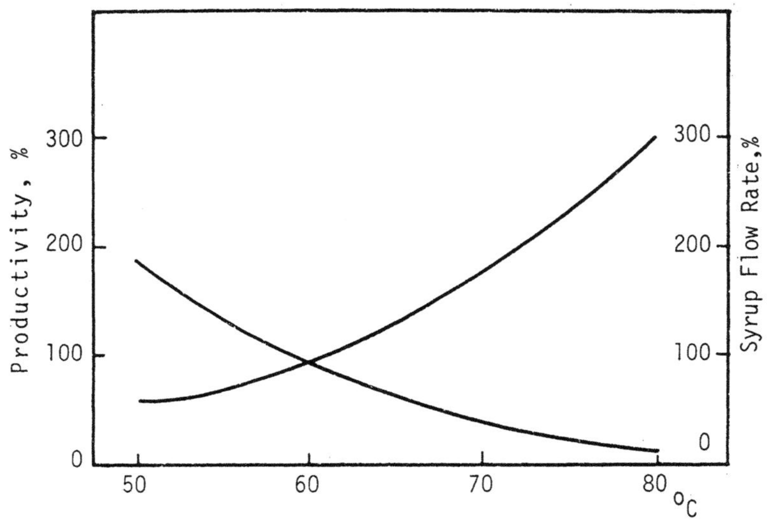

Figure 2.22 Effect of isomerization temperature. Substrate, 45% w/w glucose; pH 7.5.

2.3.8.3 Isomerization pH. The isomerization pH also affects the enzyme activity, stability and byproduct formation. Maximum activity is obtained at a pH above 8, whereas maximum stability is found between pH 7.2 and 7.5. Byproduct formation due to glucose and fructose decomposition increases at higher pH levels.

With the aim of obtaining optimal enzyme economy as well as high syrup quality, the inlet pH should be selected within the range 7.5–8.0 (measured at 25°C) depending on feed syrup purity. Feed impurities will catalyse glucose and fructose decomposition into acid by-products, thereby leading to lower outlet pH. Outlet pH readings lower than 7.0 (measured at 25°C) should be avoided. A mild buffering of the feed syrup with Na_2CO_3 or $NaHCO_3$ will facilitate pH control. (The use of Na_2CO_3 for feed pH adjustment will result in sufficient buffering in most cases).

2.3.8.4 Isomerization glucose (dextrose-DX) and fructose concentration. The feed syrup DX should be as high as possible in order to obtain the maximum isomerization rate. With low feed DX, the isomerization reaction must come closer to chemical equilibrium in order to attain a given fructose concentration, e.g. the standard 42%.

The rate of isomerization decreases with increasing fructose concentration as chemical equilibrium is approached. In consequence, any fructose content in the feed syrup (e.g. from fructose enrichment recycle streams) should be limited to a minimum.

The dry-substance content of the feed syrup should generally be

40–50%. Higher syrup concentration and higher viscosity will result in a reduced isomerization rate due to diffusion resistance in the pores of the immobilized enzyme.

2.3.9 Immobilized enzyme system

The glucose isomerases used are immobilized and granulated to a particle size between 0.3 and 1.0 mm. Mg^{2+} acts as activator and stabilizer of the enzyme, and is therefore added to the feed syrups in the form of $MgSO_4{\cdot}7H_2O$. The amount necessary depends on the presence of calcium ion, which acts as an inhibitor in the system by displacing the magnesium ion activator from the isomerase molecule. Therefore the Ca^{2+} ion content should be kept as low as possible. At a Ca^{2+} ion content in the feed syrup of 1 ppm or lower, the addition of 45 ppm Mg^{2+} (e.g. approximately 0.6 g of $MgSO_4{\cdot}7H_2O$ per l) will be sufficient. At higher concentrations of Ca^{2+} ion, a proportionate weight ratio between Mg^{2+} and Ca^{2+} ions should be provided.

Each reactor load of Sweetzyme is used for a long period of time, thereby being exposed to a very large quantity of syrup. In consequence, the accumulated effect of even low levels of food syrup impurities may lead to significant reduction in the enzyme lifetime. Some impurities may chemically inactivate the isomerase, whereas others may adsorb on to the isomerase particles, gradually blocking their active surface. In order to attain maximum enzyme lifetime, it is therefore very important to purify the feed syrup (including possible recycle streams) thoroughly before isomerization:

- The raw high-DX liquor from saccharification is filtered to remove particulate material which might clog the isomerase particles. Soluble impurities (peptides, amino acids, ash etc.) are potential inhibitors of the isomerase and must therefore be removed by carbon treatment and ion exchange. The efficiency of the carbon and ion exchange treatments should be checked by measuring the UV absorbance and conductivity of the purified feed syrup.
- Dissolved oxygen in the feed syrup increases byproduct formation, and some byproducts may inhibit the enzyme. Therefore, the oxygen content of the syrup should be minimized by vacuum deaeration after heating to isomerization temperature. Sulphite can be used as an oxygen scavenger, but should not be used in amounts exceeding approximately 100 ppm SO_2 in the feed syrup.
- Before entering the isomerization reactor the syrup should be passed through a 5–10 μm check filter.
- Some impurities (e.g. maltulose) which are difficult to remove by the above mentioned procedures may have been formed during the starch

liquefaction process. This byproduct is minimized by avoiding excessive exposure of the gelatinized starch to high temperature and pH.

The main criteria for selecting the feed syrup specifications are optimization of enzyme productivity and limitation of byproduct formation. Typical feed syrup specifications are shown in Table 2.10.

2.3.9.1 Process layout. Figure 2.23 is an example of process layout covering purification, isomerization and syrup product finishing.

2.3.9.1.1 Enzyme decay. During operation the immobilized enzyme loses activity. Most commercial enzymes show an exponential decay as a function of time as shown for SwQ in Figure 2.24, but with SwT

Table 2.10 Typical feed syrup specifications

Temperature	55–60°C
pH	7.5–8.0
Dry substance content	40–50 wt%
Glucose content	⩾95%
SO_2	0–100 ppm
Calcium ion	⩽1 ppm
$MgSO_4{\cdot}7H_2O$ (activator)	0.15–0.75 g l^{-1}
Conductivity	⩽100 μS cm^{-1}
UV absorbance (280 nm)	⩽0.5

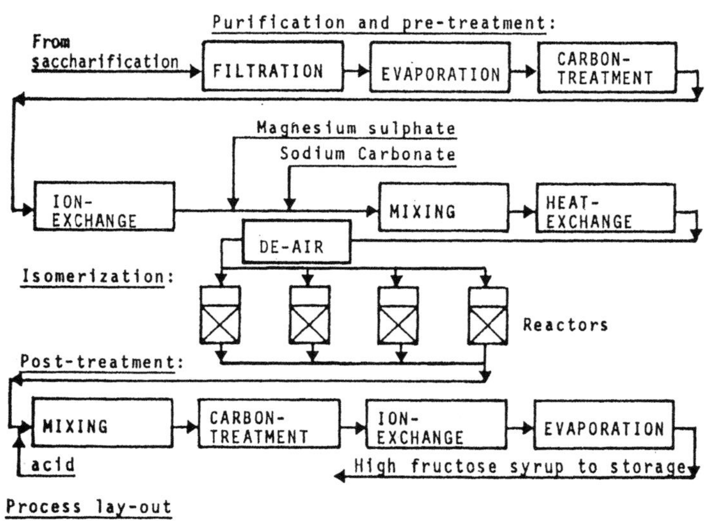

Figure 2.23 Process layout from saccharification to high fructose syrup storage.

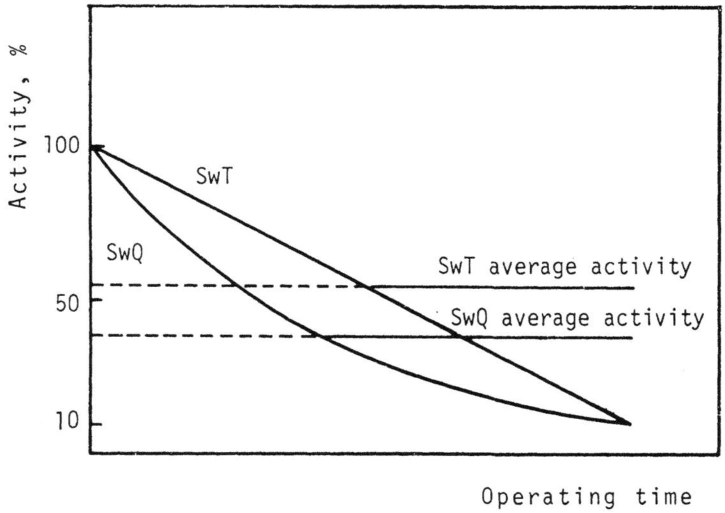

Figure 2.24 Activity (syrup flow rate) versus operating time for typical immobilized isomerases.

(Sweetzyme T is the *Streptomyces murinus* glucose isomerase from Novo Nordisk) a linear activity decay is found.

Typically, a reactor load of glucose isomerase is replaced after three half lives, i.e. when the activity has dropped to around 12.5% of the initial value. The most stable commercial glucose isomerases have half lives of around 200 days in industrial practice.

To maintain a constant fructose concentration in the product syrup, the feed flow rate is adjusted according to the actual activity of the enzyme. With only one isomerization reactor in operation, excessive variations in syrup production rate would be the result. To avoid this, several reactors containing enzymes of different ages are operated in combination.

Reactor design for glucose isomerization in the United States has been described (Blanchard and Geiger, 1984). Reactor diameters are normally between 0.6 and 1.5 m. Typical bed heights are 2–5 m. Minimum bed height:diameter ratio for one reactor is 3:1 to ensure good flow distribution. Plants producing more than 1000 t of HFCS (based on dry matter) per day use at least 20 individual reactors.

2.3.10 Economics of HFCS production

25 kg (approximately one bushel) of corn yields in the wet milling process about 0.75 kg of corn oil, 6.3 kg corn gluten feed, 1.2 kg corn gluten meal and 14 kg HFCS 55 (on a dry basis). Based upon values given in BIOTOL (1991) and in the mass balance shown in Table 2.11 the production costs have

Table 2.11 Estimated mass balance of the corn wet milling and HFCS process

Products and fractions	Mass (kg)
Corn (90% dry solids)	100
Corn oil	3.0
Corn gluten feed	25.2
Corn gluten meal	4.8
HFCS (78% dry solids)	73.0

been calculated for HFCS 55 (70–86% dry solids) as shown in Table 2.12. These data vary depending on parameters related to the actual production plant, country and production year, but they can be used as an illustration of the various costs of the different operation steps involved. A more detailed description of the development of glucose syrups, their science and technology and progression as the basis of a highly complex, biotechnological based industry is described by Dziedzic and Kearsley (1984).

2.3.11 Other sweeteners

Novel developments in enzymic modification of starch are among other things concerned with production of cyclodextrins. The cyclodextrins are cyclic molecules consisting of 6–8 glucose units linked together by α-1,4-bonds. They can bind other compounds in their cavity and thereby stabilize, solubilize or precipitate compounds (Pszczola, 1988). Cyclodextrins are produced during liquefaction of starch by use of the class of enzymes called cyclodextrin glycosyl transferases (CGTase) (Nielsen, 1991).

Production of fructose from inulin is one use of enzymes for production of sweeteners and the use of invertase is another alternative available to alter the sweetness and functional properties of sweeteners.

2.3.11.1 Enzymic hydrolysis of inulin in the production of fructose. Fructose is the sweetest naturally occurring nutritive sweetener. An alternative source of fructose is inulin, which occurs in the roots of plants such as chicory and Jerusalem artichoke. Inulin is a linear, β-2,1-linked fructose polymer initiated by a glucose unit. The fructose content varies depending on the insulin source, but is typically 90%, corresponding to an average chain length of 10 units.

Inulin may be saccharified by use of a mixture of *exo*-inulinase (EC 3.2.1.80) and *endo*-inulinase (EC 3.2.1.7), e.g. the product Fructozyme^{TM} obtained from a selected strain of *Aspergillus niger* and produced by Novo Nordisk (Zittan, 1981; Application sheet, 1993).

Table 2.12 Production costs of HFCS 55 in US$ per 100 kg of HFCS (70–86% dry solids; average = 78%)

Process step	Values in US$ per 100 kg of HFCS (70–86% dry solids; average = 78%)
Sales values	
Value of byproducts	11
Value of HFCS 55	28–40 (1989)
Costs of HFCS production	
Raw material	
- Corn (137 kg)	(10.70)[a]
- Net starch costs	8.80
Capital depreciation	2.30
Labour (2.5 men)	2.00
Milling and feed preparation	7.00
Refinery utilities	1.00
Liquefaction enzymes	0.17
Saccharification enzymes	0.35
Clarification filter aid	0.12
Carbon treatment	
- dextrose	0.30
- HFCS	0.20
Ion exchange	
- resins	0.23
- chemicals	0.35
Isomerization	
- enzyme	0.38
- chemicals	0.03
Fractionation	
- resin	0.12
- evaporation	0.16
Final evaporation	0.32
Total production cost	23.85

[a] Not used for calculation

Inulin is extracted from sliced plant material by diffusion at elevated temperatures. The raw juice can be purified by liming and carbonation, similar to sugar beet processing, and then decolorized by carbon. A dry substance content in the raw juice of 15–25% is optimal for Fructozyme. A degree of hydrolysis above 98% is obtained within 12 to 48 h depending on enzyme dosage. The fructose syrup produced can be used as is or for blending for the production of 55% fructose syrup. Alternatively, the purified syrup can be used for production of crystalline fructose (Application sheet, 1993).

2.3.11.2 Enzymatic hydrolysis of sucrose. Invertase hydrolyses sucrose to glucose and fructose. Invertase is used in the confectionery industry to convert the easily crystallized sucrose into the less easily crystallized glucose–fructose mixture. 'After Eight Mints', for example, have invertase in their centres. Thereby the hard sucrose core (coated with chocolate) is turned to the soft centre, which finally we eat (Bains, 1993).

2.4 The future

Value added contributions hopefully do not stop with the achievements mentioned above. The obvious attributes of sweeteners from starch using enzyme techniques offer a complex range of options.

Refined starch sweeteners have proved so important to the food industry that for example since 1985 the Americans have used more corn products than sucrose for their nutritive sweetener needs (Munro, 1994). The wet miller's value added contributions do not stop with these achievements. The industry maintains its emphasis on science and creation of new products. During recent years introduction of a wide range of value added food ingredients that are still being explored have been seen; for example fat replacers from starch, cyclodextrins for encapsulation, crystalline fructose and new modified starches for microwave applications.

2.4.1 Speciality sweeteners

Sugar alcohols (polyols) like sorbitol and mannitol have been available for more than 50 years and the food industry originally used them to sweeten foods for diabetics, because rapid glycaemic response is avoided. With the growing demand for non-cariogenic and low calorie foods, the available range of polyols had been extended in order to provide process flexibility and improved organoleptic properties.

2.4.2 Polyols

Polyols like sorbitol, mannitol and xylitol are found in low concentrations in a number of fruits, vegetables, cereals etc. Extraction is therefore not an economical way to produce these products. Industrial production of polyols is achieved by catalytic hydrogenation of easily accessible carbohydrates like the starch hydrolysates described. Dextrose and glucose syrups are used for sorbitol products, high maltose syrups for maltitol, xylose for xylitol etc. These products are available as crystalline material or as concentrated solutions.

2.5 Conclusions

Glucose syrups as starch-derived bulk sweeteners have been shown to be much more than just sweeteners. Many food products and chemicals are derived from the utilization of starch as a raw material for enzymic conversions. Thus it has been demonstrated that the major steps in conversion of starch, liquefaction, saccharification and isomerization, may lead to many valuable products, when processing and enzymology go hand-in-hand. In simple terms, the further a starch processor proceeds, the more diverse products can be obtained.

Acknowledgements

Most data have been taken from the information material from Novo Nordisk A/S. I am also grateful to Mr Erik Andersen and Mr Tommy Rex Christensen, starch industry product managers at Novo Nordisk, for their help and advice during the preparation of the manuscript.

References

Application sheet (1993) *Use of Fructozyme™ in the Production of Fructose from Inulin.* Enzyme Process Division of Novo Nordisk A/S, Bagsvaerd, Denmark.

Application sheet (1990) *Use of Termamyl® for Starch Liquefaction.* Enzyme Process Division of Novo Nordisk A/S, Bagsvaerd, Denmark.

Bains, W. (1993) *Biotechnology from A to Z*, Oxford University Press.

BIOTOL (1991) *Biotechnological Innovations in Food Processing*, Butterworth-Heinemann, Oxford.

Blanchard, P.H. and Geiger, E.O. (1984) *Sugar Technol. Rev.* **11**, 1–94.

Chouvel, H., Chay, P.B. and Cheftel, J.-C. (1983) Enzymatic hydrolysis of starch and cereal flours at intermediate moisture contents in a continuous extrusion-reactor. *Lebensm.-Wissen.-Technol.* **16**, 346–353.

Dziedzic, S.Z. and Kearsley, M.W. (1984) *Glucose Syrups: Science and Technology*, Elsevier Applied Science, London and New York.

Jørgensen, O.B., Karlsen, L.G., Nielsen, N.B., Pedersen, S. and Rugh, S. (1988) A new immobilized glucose isomerase with high productivity produced by a strain of *Streptomyces murinus*. *Starch/Stärke* **40**, 307–313.

Konieczny-Janda, G. and Richter, G. (1991) Progress in the enzymatic saccharification of wheat starch. *Starch/Stärke* **43**, 308–315.

Lii, Cheng-Yi and Chang, Yung-Ho. (1991) Study of starch in Taiwan. *Food Rev. Internat.* **7**, 185–203.

Munro, E.M. (1994) Corn refining: a classic value-added success story. *Cereal Foods World* **39**, 552–554.

Nielsen, H.K. (1991) Novel bacteriolytic enzymes and cyclodextrin glycosyl transferase for the food industry. *Food Technol.* **45**, 102–104.

Norman, B.E. (1982) A novel debranching enzyme for application in the glucose syrup industry. *Starch/Stärke* **34**, 340–346.

Pszczola, D.E. (1988) Production and potential food application of cyclodextrins. *Food Technol.* **42**, 96.

Reichelt, J.R. (1983) *'Starch' in Industrial Enzymology* (ed. T. Godfrey and J. Reichelt), Macmillan, England.

Rosendal, P., Nielsen, B.H. and Lange, N.K. (1979) Stability of bacterial alpha-amylase in the starch liquefaction process. *Starch/Stärke* **31**, 368–372.

Sims, K.A. and Chernyan M. (1992a) Hydrolysis of liquefied corn starch in a membrane reactor. *Biotechnol. Bioeng.* **39**, 960–967.

Sims, K.A. and Cheryan M. (1992b) Continuous saccharification of corn starch in a membrane reactor. *Starch/Stärke* **44**, 341–346.

Tewari, Y.B. and Goldberg, R.N. (1984) *J. Solution Chem.* **13**, 523–547.

Zittan, L. (1981), Enzymatic Hydrolysis of Inulin – An Alternative Way to Fructose Production. *Starch/Stärke* **33**, 373–377.

3 Maltodextrins

J.F. KENNEDY, C.J. KNILL and D.W. TAYLOR

3.1 Introduction

Commercial starch hydrolysates are classified on the basis of dextrose equivalent (DE). Maltodextrins are by definition starch hydrolysates (non-sweet saccharide polymers) that consist of α-D-glucose units linked primarily by (1→4) glycosidic linkages with a DE of less than 20, and a general formula of $[(C_6H_{10}O_5)_nH_2O]$. DE is defined as the percentage of reducing sugar in a syrup calculated as dextrose on a dry weight basis. The definition of maltodextrins can thus be taken further as those materials having a dextrose equivalent of between 3 and 20. They, therefore, essentially bridge, in terms of molecular size, the gap between starch and sugar (Morris, 1984). The DE reflects reducing power, and therefore indicates stability and functionality. Starch hydrolysates with a DE greater than 20 are designated as various kinds of syrups depending on their source (e.g. corn syrups). Figure 3.1 shows the structural components of a maltodextrin mixture.

Starch hydrolysates are increasing in commercial importance due to their unusual properties. In general maltodextrins are fully soluble carbohydrates of low bulk density, they can be metabolised in a similar way to starch and in some applications are therefore suitable for diabetics (Anon, undated (a)). Interestingly maltodextrins have very little or no sweetness and a bland, not starchy flavour that does not mask other flavours (Morris, 1984; Anon, undated (a)). Due to these properties maltodextrins are finding a considerable number of applications within the food industry, particularly in convenience and processed foods.

3.2 Maltodextrin production

The production of maltodextrins is, by definition, achieved by hydrolysis of starch down to glucose polymers with an average chain length of 5–10 glucose units/molecule. Theoretically they can be produced by controlled hydrolysis by either enzymic (α-amylase) or acidic (usually hydrochloric acid) means (Anon, 1989a), however in practice acid hydrolysis produces too much free glucose and maltodextrins thus produced have a strong tendency to retrograde, resulting in the formation of hazy solutions.

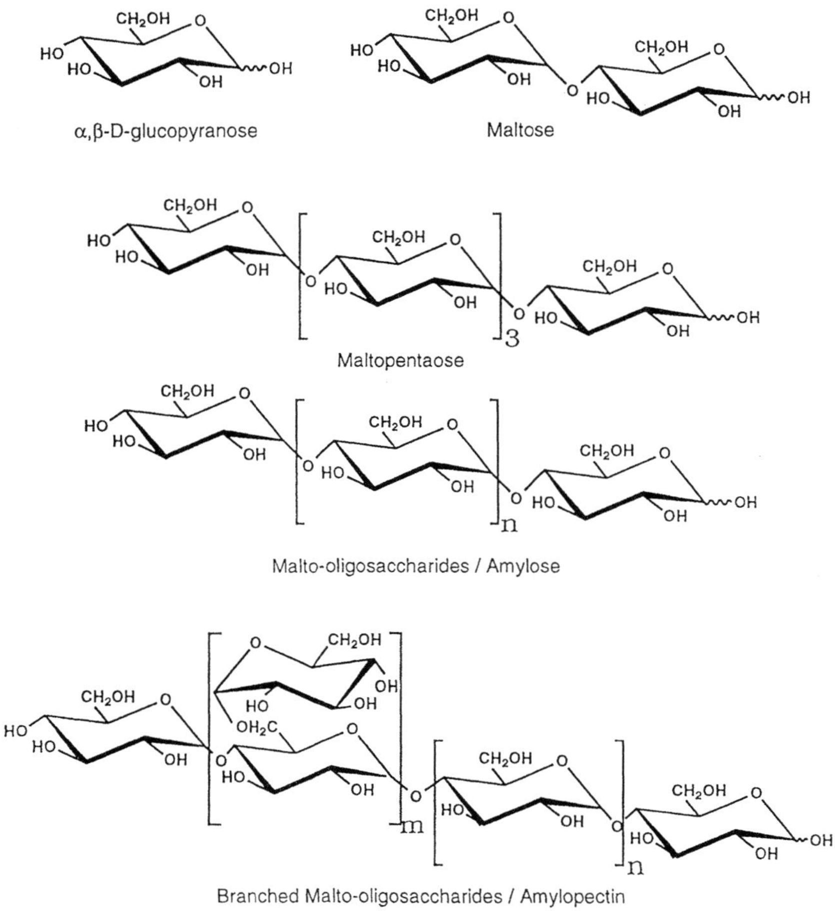

Figure 3.1 Main structural components of a maltodextrin mixture.

Linear starch fragments in low DE acid hydrolysates large enough to reassociate form insoluble aggregates causing hazy solutions which are undesirable for some applications. Therefore on a commercial level maltodextrins are invariably prepared from starch by controlled enzymic hydrolysis (Morehouse *et al.*, 1972a, b). A starch slurry is initially liquefied by heating to 70–90°C at neutral pH in the presence of a bacterial α-amylase to a DE of 2–15. The liquid starch hydrolysate is then autoclaved at 110–115°C to gelatinise completely any remaining insoluble starch and, on cooling, subjected to further enzymic treatment to reach the desired DE. Some hydrolysis schemes however, employ both acid- and enzyme-catalysed hydrolysis. Initial acid-catalysed hydrolysis of a starch slurry to a DE of 5–15 followed by neutralisation and further hydrolysis with a bacterial α-amylase such as that obtained from *Bacillus subtilis* or *Bacillus mesentericus* is a recommended procedure to produce maltodextrins that are haze free and exhibit no retrogradation upon storage (Armbruster and Harjes, 1969, 1971). Such systems overcome the problem of hazy solutions and result in maltodextrins with low hygroscopicity and high water solubility. The source and variety of the starch is unimportant for these processes.

Following hydrolysis, the pH of the crude maltodextrin solution is adjusted to about 4.5 and the solution filtered to remove small amounts of fibre, lipid and protein. Filtration can prove difficult due to the high viscosities of maltodextrin solutions and this processing stage can lead to substantial losses in yield. The clarified solution is then further refined by one or more of the procedures used for corn syrups, such as carbon refining. The refined solution is then concentrated in vacuum evaporators to give finished syrups containing about 75% solids, or more regularly, is spray dried to a white powder containing some 3–5% moisture.

The dextrinisation reaction is allowed to progress until the required DE product is obtained. To give some idea of time scale, after 1 h the DE is 10–12, by 2 h it is 15–16. Low DE value maltodextrins will tend to retrograde in solutions whereas those with higher DE values will form less viscous solutions and will exhibit increasing sweetness. The component breakdown of two maltodextrin mixtures (DE 15 and 20) is shown in Table 3.1.

Two types of maltodextrin are in commercial use: those ranging from about 10–14 DE and those ranging from about 15–19 DE. The compositions of these products depend not only on DE, but also on the method of hydrolysis employed in their manufacture. The saccharide component profiles of maltodextrins obtained by acid-catalysed hydrolysis are somewhat different from those obtained by enzyme- or acid/enzyme-catalysed hydrolysis, even though they can possess the same DE. Acid hydrolysates tend to contain greater proportions of high molecular weight dextrins, which retrograde easily.

Gelling maltodextrins are obtained by hydrolysis of potato starch with bacterial α-amylase. At concentrations above 10% (w/w) and temperatures below 30°C their aqueous solutions solidify to give thermally reversible gels. Melting takes place as a function of concentration between 45°C and 85°C. Maltodextrin gels are two-phase systems resulting from

Table 3.1 Composition of two maltodextrin mixtures (DE 15 and 20)

DE	Glucose (%)	Maltose (%)	Maltotriose (%)	Higher saccharides (%)
15	0.6	4.0	7.0	88.4
20	0.8	5.5	11.0	82.7

mutual interactions between amylose molecules and sufficiently long amylopectin chains resulting in the formation of a mixed gel structure (Schierbaum *et al.*, 1986; Vorweg *et al.*, 1988).

3.3 Properties of maltodextrins

The varying properties of maltodextrins are indicated by the DE and the degree of polymerisation (DP), which change with the degree of hydrolysis and enzymic treatment (Morris, 1984; Ziesenitz and Siebert, 1987). The solubility of maltodextrins varies with DE and with the method of hydrolysis. Enzyme-hydrolysed products usually contain lower concentrations of the high molecular weight saccharides and are thus more water soluble than acid-hydrolysed products of equivalent DE. Maltodextrins are relatively non-hygroscopic compared with corn syrups, and those having the lowest DE exhibit the least tendency to absorb atmospheric water (Morris, 1984).

The high viscosity of maltodextrins, an important property in many applications, is due to high levels of high molecular weight saccharides (Anon, undated (b)). Since maltodextrins exhibit virtually no sweetness, their main contribution is a 'bodying effect' resulting from their relatively high viscosity. The varying DE/bulk density give maltodextrins their varying characteristics and functionality. For example, the high DE maltodextrins have solubility, bulking and bodying characteristics more similar to the corn syrup sweeteners. The low DE maltodextrins have the binding properties of starch and can function more effectively as fat binders than high DE maltodextrins.

Low DE maltodextrins are effective in preventing formation of coarse crystals (Anon, undated (a)). The variety of sugar polymers in maltodextrins prevents the formation of large, gritty crystals as sugar mixtures will not crystallise as readily as pure compounds. The fine particle size of maltodextrins contributes to a smooth, creamy mouthfeel in many food products (Anon, undated (a)).

3.4 Analysis of maltodextrins

Chromatography is one of the best techniques for the characterisation of oligo- and polysaccharides. High performance liquid chromatography (HPLC) is an interesting technique for the investigation of oligosaccharides. Since its inception in the mid 1970s the number of methods and types of packing materials have expanded rapidly. Much of the earlier work on oligosaccharide analysis has been reviewed (Ben-Bassat and Grushka,

1991) and the theory, separation modes and instrumentation described (Macrae, 1982).

HPLC is a technique by which molecules in solution are separated (fractionated) according to differences in their molecular sizes, ionic properties or affinity for a column packing material. For HPLC high pressures must be used in order to permit a sufficient flow rate of the mobile phase. The separation time is an important factor in modern HPLC. The main objective is to achieve the best separation in the minimum time. The acronym HPLC can also stand for high pressure LC, but is now usually understood to mean high performance LC as the use of pressure is not the goal but merely an auxiliary means of achieving good separations (Lim, 1986).

The three main techniques used in HPLC are ion exchange chromatography (IEC), reversed phase chromatography (RPC) and affinity chromatography. IEC is one of the most important separation processes in biochemistry and other related areas, since many substances of biological origin are water-soluble ionic compounds. Bonded silicas and bonded glass with ionic groups on their surface are used in such HPLC techniques as the stationary phase separation media. Ionic solute molecules are attracted to the stationary phase ionic groups of the opposite charge and during elution the retarded substances are reversibly exchanged for ions of the same charge. Anion exchange HPLC has been recently developed and has exceptional resolving power for complex oligosaccharides. Such analyses are carried out at high pH coupled with pulsed amperometric detection (PAD), allowing separation of oligo- and polysaccharides up to DP$\geqslant$50 (Koizumi, *et al.*, 1989). The separation depends on the molecular size, sugar composition and type of linkages between the monosaccharide units.

RPC has become a widely used HPLC separation mode. In RPC the mobile phase is polar (aqueous solutions) and the stationary phase is non-polar (organic polmers or hydrocarbon chains bonded on silica). This arrangement of the contact area for both phases is the reverse of that in both classical adsorption chromatography and 'normal' partition chromatography (Szepesi, 1992). The use of fully acetylated oligosaccharides overcomes problems of solubility and has resulted in the fractionation of malto-oligosaccharides up to DP30 in about 150 min using RPC (Wells and Lester, 1979).

By far the most frequently used systems for separation of oligosaccharides are those using chemically bonded phases which fractionate materials on the basis of their relative affinities for the mobile phase and the bonded phase. The most important type of columns are those containing the aminopropyl bonded phase (Schwarzenbach, 1976) which are regarded by some workers as the ultimate in column technology for oligosaccharide analysis (Folkes and Taylor, 1982). Non-polar organic solvents or an aqueous/organic mixture (such as aqueous methanol or acetonitrile) are usually used as eluent. Separation of a series of oligomers from a starch hydrolysate (such as a maltodextrin mixture) can readily be achieved with up to DP 10 being separated in 15–20 minutes using acetonitrile/water eluents containing 35–40% water (see Figures 3.2(a) and (b)). However very high molecular weight materials cannot be analysed due to prolonged retention times and solubility problems in the acetonitrile/water eluents. Therefore for a full analysis separation gel permeation or ion exchange is also required.

Compounds that do not interact with the column packing media are simply filtered through the interstitial volume between particles and emerge from the column in the dead volume (or void volume, V_0). Solvent elution in HPLC is either isocratic elution (constant concentration of eluent) or gradient elution (predetermined variation of solvent composition during the run). Gradient elution is a must for many HPLC applications.

Detection of oligosaccharides eluting from HPLC columns is the biggest challenge and the weakest link in the analytical chain. For underivatised

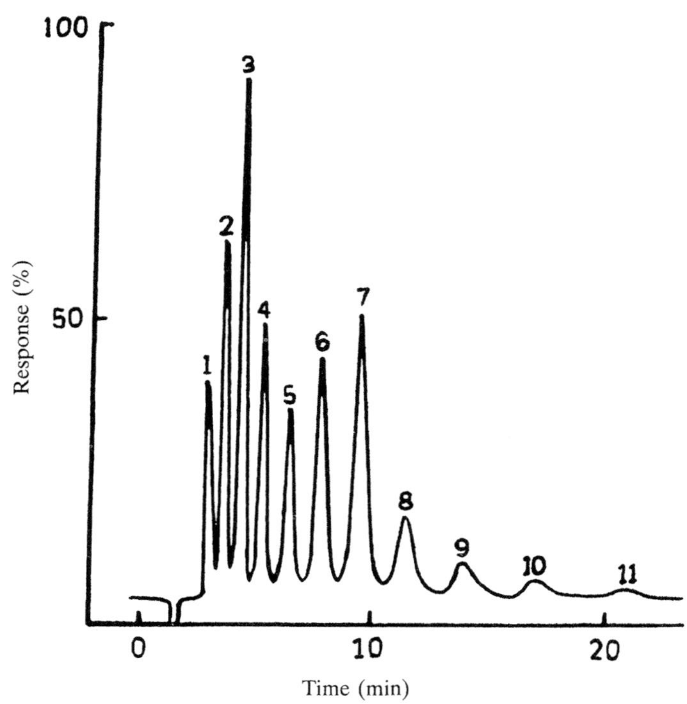

(a)

Figure 3.2 Separation of a starch hydrolysate (a) on a Spherisorb S5 NH_2 column; (b) on an amine dynamically modified silica column (5 μm particle size, 200 × 8 mm i.d. with 50 × 5 mm presaturation column). 1, 2, 3, etc., refer to the DP of the oligosaccharides.

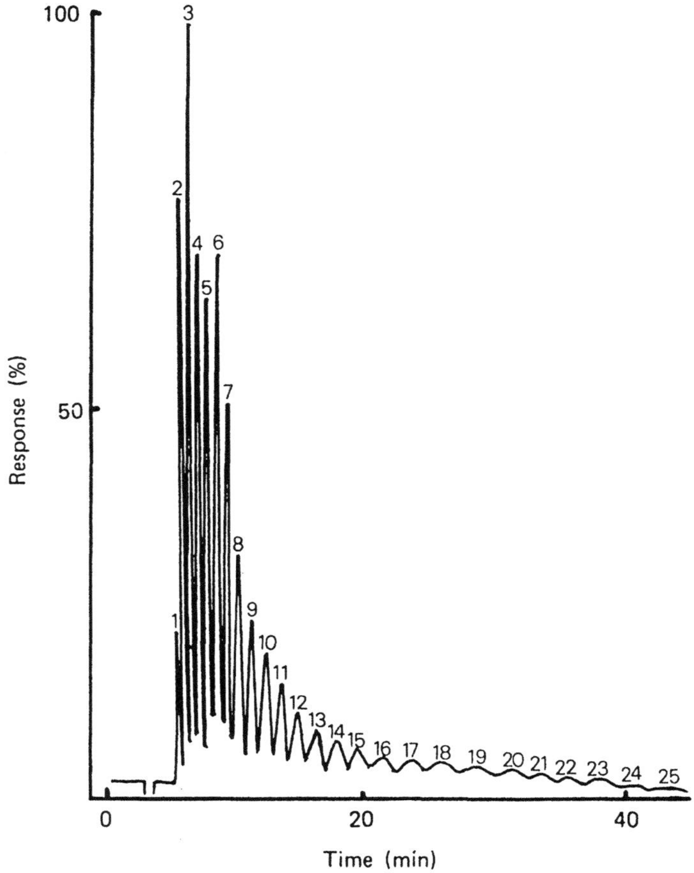

(b)

Figure 3.2 *continued*

oligosaccharides the most common detectors are refractive index (RI), UV and electrochemical detectors. Non-specific detectors such as the RI detector are used routinely (Scott and Hatina, 1988; Derler *et al.*, 1988). PAD detectors have been considerably developed in the last few years, coupled in particular with anion exchange (Hernandez *et al.*, 1990) and affinity HPLC (Wang *et al.*, 1992). A complete study of linear and branched glucose oligomers using anion exchange HPLC and PAD has recently been performed to look at variations in detector response at different degrees of polymerisation (Ammeraal *et al.*, 1991). A considerable amount of work outlining the chromatographic characterisation of starch oligosaccharides has been published previously (Heyraud and Rinaudo, 1991).

Gel permeation chromatography (GPC), or size exclusion chromatography (SEC) as it is sometimes known, is the most frequently applied method in this area, and is a technique by which molecules in solution are separated (fractioned) according to differences in their molecular sizes (or more specifically their hydrodynamic volume) as they pass through a column packed with one of the many chromatographic media available (in this case a gel) (Hunt and Holding, 1989). A gel is a heterogeneous phase system in which a continuous liquid phase, normally aqueous in nature, is contained within the pores of a continuous solid phase, the gel matrix. The pores have a specifically controlled range of sizes and the matrix is chosen for its chemical and physical stability and inertness (lack of adsorptive properties). Gels may be formed from polymers by cross linking to form a three-dimensional network; for example 'BioGel' is formed by crosslinking polyacrylamide and 'Sephadex' by crosslinking dextran.

The pores in the gel matrix which are filled with the liquid phase are usually comparable in size to the molecules of interest for separation. In a gel filtration column gel particles in bead form are packed to form a separation bed through which a buffer solution, the eluent, is passed. Sample molecules which are to be separated are added as a zone to the top of the bed. The sample zone moves down the bed as eluent is added to the top. The small molecules which diffuse into the gel beads are delayed in their progress down the column compared with the large molecules which cannot diffuse into the gel and move continuously down the column in the flowing eluent (and are said to be excluded, since their size is greater than the exclusion limit of the gel). The large molecules thus leave the column first followed by the smaller molecules in order of their sizes. The eluent is usually a buffer solution with a well defined pH and ionic composition chosen to preserve the structure and biological activity of the substances of interest.

GPC has occupied a key position in the purification of thousands of enzymes (Chaplin and Bucke, 1990), polysaccharides (Kennedy and White, 1983), proteins (Simpson, 1993) and other biological macromolecules. It is particularly useful in the separation of many different carbohydrates, and using BioGel P2 polyacrylamide gel as the separation media mono- and oligosaccharides can be successfully separated, the exclusion limit of the gel being 2000. The gel permeation chromatography of maltodextrins has been investigated and BioGel P2 has been shown to give a good separation based on the hydrodynamic volume of the carbohydrate, provided that column temperature is maintained constant at greater than 65°C to suppress the loose adsorption that exists on this polyacrylamide based matrix (see Figure 3.3 for column set up). When linked to an automated on-line cysteine–sulphuric acid assay system (Figure 3.4) to monitor the total carbohydrate content of the material coming off the column (Barker *et al.*, 1968), maltodextrin samples can be

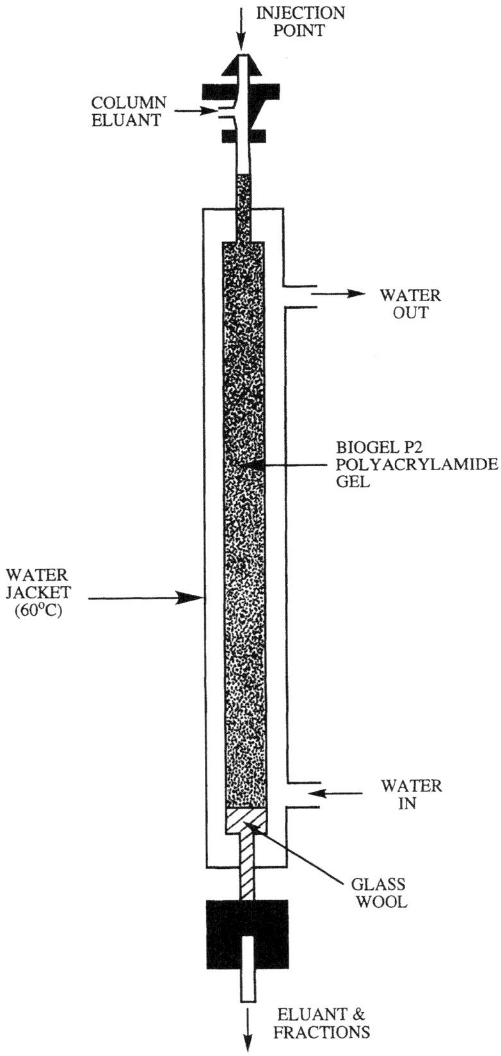

Figure 3.3 BioGel P2 gel permeation chromatography (GPC) system.

analysed routinely to give information on their molecular weight distribution in terms of degree of polymerisation (Figures 3.5(a) and (b)). Peak areas can easily be quantified to give the percentage of each oligomer in the total material (Table 3.2).

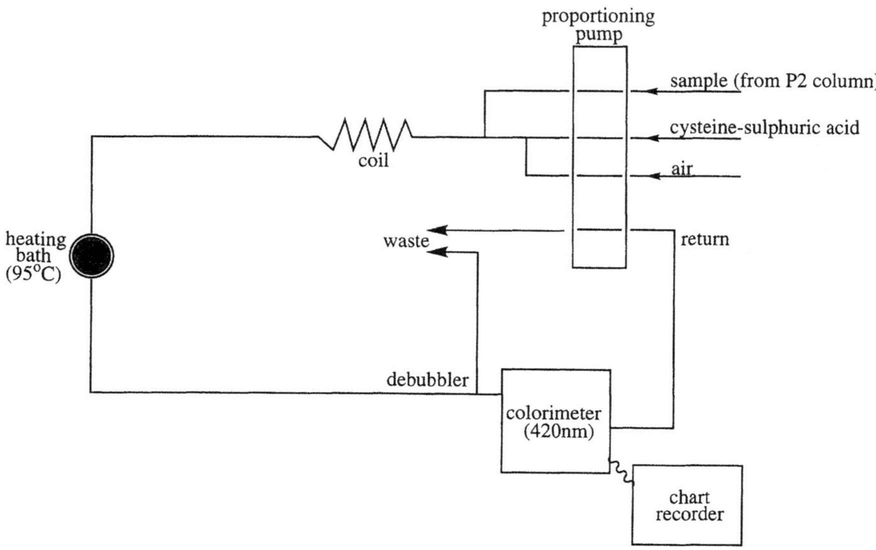

Figure 3.4 Automated on-line L-cysteine–sulphuric acid assay system for total neutral carbohydrate determination.

Table 3.2 Molecular weight profile of a commercial maltodextrin mixture (in terms of degree of polymerisation, DP, see Figure 3.5(a) for component profile); (IMW and HMW refer to intermediate and high molecular weight respectively).

Peak	Area (%)
DP1	0.6
DP2	2.4
DP3	4.0
DP4	8.6
DP5	11.1
DP6	12.3
DP7	8.4
DP8	3.7
IMW	15.3
HMW	33.6

3.5 Applications of maltodextrins in food and related areas

Maltodextrin uses are widespread throughout the food industry and they perform multifaceted functions in food systems; a wide range of products are available in powdered and syrup forms. The functionality of each type varies with extent of starch degradation. If a glucose syrup is spray dried to produce a soluble white powder it gives a product which is convenient and easy to handle. The powder's characteristics are, to a certain extent, dependent upon its DE, but in all cases the maltodextrin will be soluble, have a low bulk density and low sweetness.

Batches with a higher DE are more sweet and more hygroscopic than their low DE counterparts. The properties and cost of maltodextrins make them ideally suited to application within the food industry, where their uses include:

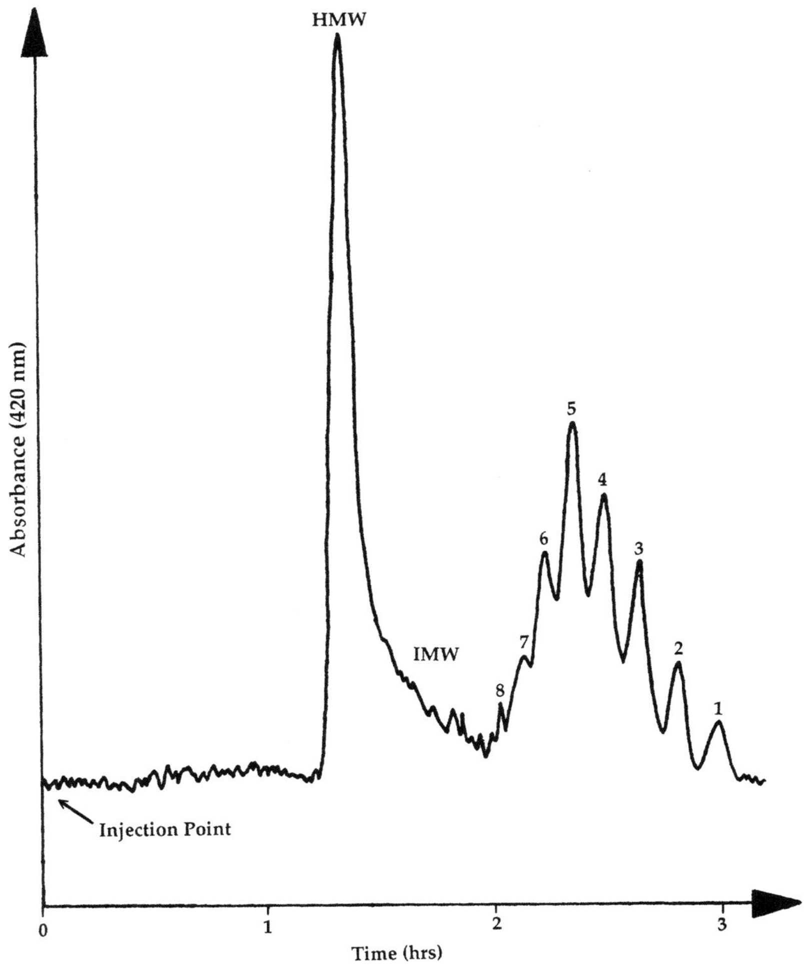

Figure 3.5(a) P2-GPC chromatogram of a commercial maltodextrin mixture (25 μl injection of a 1% w/v solution; column eluent 0.1 M NaCl; column 60 × 1 cm i.d.; flow rate 0.2 ml min^{-1}; 1, 2, 3, etc., refer to the DP of the oligosaccharides; IMW and HMW refer to intermediate and high molecular weight, respectively).

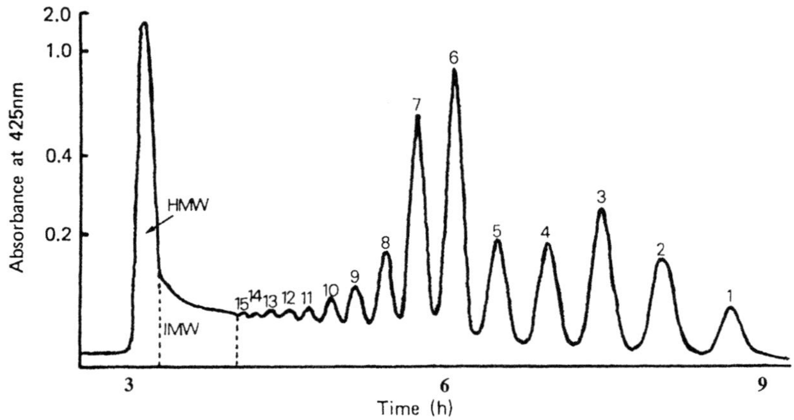

Figure 3.5(b) P2-GPC chromatogram of a starch hydrolysate (column eluent 0.1 M NaCl; column 140 × 2.7 cm i.d.; flow rate 1 ml min^{-1}; 1, 2, 3, etc., refer to the DP of the oligosaccharides; IMW refers to material in the range DP15 to DP20; HMW refers to material above DP20).

(i) as bulking/filling/carrier agents,
(ii) as sweetness reducers,
(iii) other uses, e.g. oxygen barrier and film forming.

3.5.1 Bulking/filling/carrying

Due to their low sweetness and high solubility maltodextrins are ideally suited for blending into powdered formulations. The purposes for which they have been used include:

- to save costs by replacing (or partially replacing) a higher cost material. Examples are lactose/milk powder in certain applications such as tablet manufacture, albumin in marshmallows etc.
- as a non-flavoured dilutant for concentrated ingredients such as flavourings and flavour enhancers
- to reduce loss of volume on storage or transit
- to absorb oil/fats
- to aid dispersion
- to give a bland taste
- to increase nutritive content
- to increase solubility.

3.5.2 Sweetness reducers

Maltodextrin addition is by far the most effective method for reducing sweetness without altering properties currently available to the food/

beverage industry. Many products can be made more palatable by lowering their level of sweetness without reducing their total solids content. This is of use in foods and drinks containing high sugar concentrations (for example confectionery, preserves, fillings and fondants). Additionally in recent years this utilisation has spread to isotonic energy supplement drinks specifically designed for those in a sporting or active environment. The success of these energy replacement drinks may, however, have a great deal more to do with fashion and successful advertising policy than science.

3.6 Maltodextrin fat replacers

Maltodextrins, particularly low DE products serve as fat and oil replacers. Maltodextrin substituted directly into a food product provides 4 kcal of energy per gram of solid material (Altschul, 1989). However when maltodextrins are being used as oil or fat replacers they are generally utilised in a 1:3 maltodextrin:water ratio thus giving a solution or gel (depending on DE) which provides only 1 kcal g^{-1}. Several commercial products are already being marketed as fat replacers and many more are in development. Some of the existing fat replacers will now be described in order to provide the reader with some idea of the variety and potential of maltodextrins in this area, and to give an idea of the suitability of these products for the task at hand.

A potato starch, enzymically derived, maltodextrin known as Paselli SA2 has been developed (Anon, 1988). This fully digestible product has been successfully used to replace in excess of 50% fat in a variety of products (Kaper and Gruppen, 1987; Anon, 1990a). It has a DE of approximately 3 (therefore it is technically not a true maltodextrin) and a slightly acidic pH. Under the correct conditions and concentrations it forms a smooth gel with a fat-like texture and little or no taste. The calorific content of this gel is ca. 1 kcal g^{-1} and its only major drawback is it hygroscopicity which means that it does not store well for more than approximately 12 months.

Another commercially available maltodextrin is Amalean I which is a modified high amylose food starch which has been used to replace 100% of fat in some cases, resulting in a calorific reduction of some 96%! (LaBell and Duxbury, 1990). It has been used as a dry material or as a precooked paste. A suitable gel for fat replacement can be achieved by heating a paste or slurry of suitable proportions (8% maltodextrin is often sufficient) to ca. 90°C for 3 min.

Stellar™ is a recent but important addition to the field and is a fat mimicking microparticulate corn starch gel that restructures and immobilises water (Pszczola, 1991). This tightly held water is unable to migrate in, for example, bakery products and this is thought to be responsible for its slowing of the staling process (Anon, 1991). Using the product, a low fat (2.2%) Danish pasty can be made to stay soft for up to 35 days.

3.7 Specific examples of maltodextrin food applications

Baked goods – Maltodextrins can be used in bakery goods as a gram for gram replacement for either sucrose or fat. Examples include Paselli SA2 used in cakes, muffins and cookies as a fat replacer and Amalean I used in reduced fat products such as cakes and cheesecakes.
Frozen desserts – Replacement of sugar often requires a combination of two or more bulking agents in order to directly duplicate the functional properties of that sugar (e.g. freeze point depression, ice crystal control and structure/bulk contribution). Frozen desserts use maltodextrin because its water holding capacity combined with relatively low molecular weight give it freezing point depressant qualities.
Low calorie dry mix – A wide variety of low calorie mixes have been produced with the aid of artificial sweeteners, including hot and cold beverage mixes. Once again the maltodextrin is used to replace the bulking characteristics of the original sweetener since many sugar substitutes have intense sweetness, and therefore are only required in small quantities.
Margarine spreads – Maltodextrins are used in reduced calorie margarine spreads, with a typical reduced calorie table spread containing just 4.5 kcal g^{-1} as opposed to 7 kcal g^{-1} for a conventional margarine whilst retaining similar properties (Anon, 1990b).
Reduced fat meat products – Paselli SA2 has been used to decrease the fat levels of meat products. Maltodextrins are also finding use in reduced fat soups and gravy mixes (Anon, 1989b, undated(c)), their purpose again being to reduce fat by direct replacement.
Salad dressing – Maltodextrins have successfully been used in the production of reduced fat salad dressings (Klis, 1984).

3.8 Cyclodextrins

The ‘molecular encapsulation’ of many compounds can be performed with cyclodextrins. This operation often advantageously modifies the various physical and chemical properties of the encapsulated molecules. This method is simpler and cheaper than most other methods of encapsulation. The molecular structure of cyclodextrins renders them able to include other molecules, as a result of which powder-like crystalline inclusion complexes are formed.

The molecular dimensions of the three most significant cyclodextrins, α-, β- and γ-cyclodextrins, built up of 6, 7 and 8 glucopyranose units respectively are shown in Figure 3.6 (Zsadon *et al.*, 1978, 1979). As a consequence of the C_1 conformation of the glucopyranose units, all secondary-OH groups are located on one side of the torus-like cyclodextrin molecule, while all primary-OH groups are on the other side. The 'lining' of the internal cavity is formed by H-atoms and glycosidic oxygen bridge atoms, therefore this surface is slightly apolar. The structure of α-cyclodextrin is shown in Figure 3.7 (French, 1987). *β*-Cyclodextrin has a slightly sweet taste, is odourless, and is fairly stable against acid hydrolysis (Szejtli and Budai, 1976). In alkaline media, like other non-reducing oligo- and polysaccharides it is also rather stable.

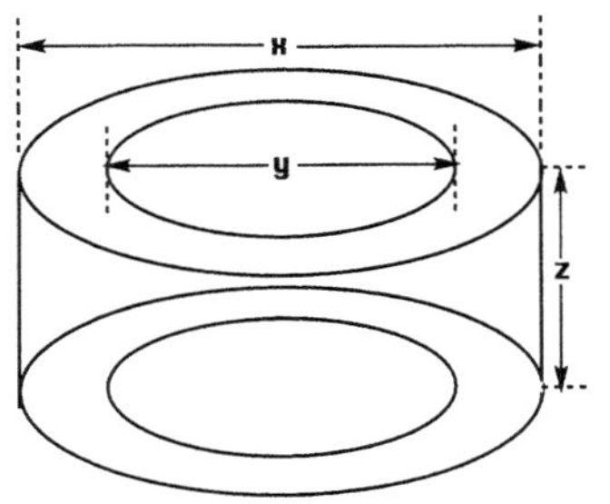

	x	y	z
α	13.7 Å	5.7 Å	7.8 Å
β	15.3 Å	7.8 Å	7.8 Å
γ	16.9 Å	9.5 Å	7.8 Å

Figure 3.6 Molecular dimensions of α-, *β*- and γ-cyclodextrins (Simpson, 1993; Barker *et al.*, 1968).

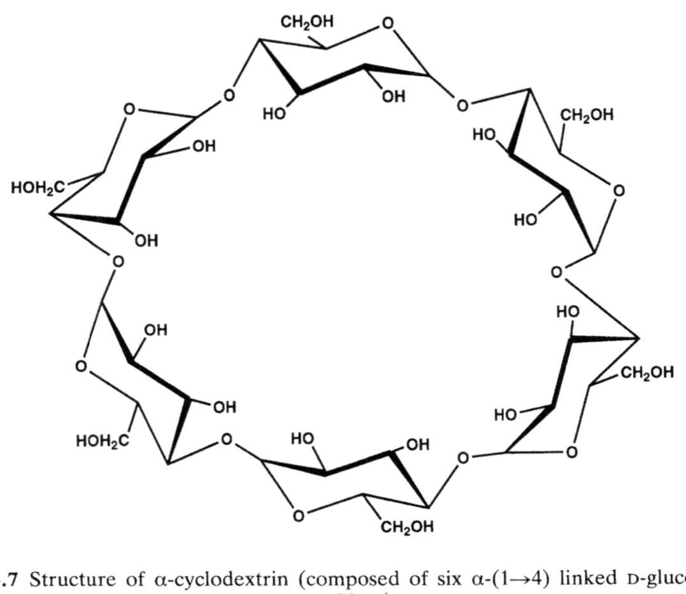

Figure 3.7 Structure of α-cyclodextrin (composed of six α-(1→4) linked D-glucopyranosyl residues).

Molecules, or functional groups of molecules, having molecular dimensions that correspond to the cyclodextrin cavity, being less hydrophilic than water, can be included in the cyclodextrin cavity if both components are dissolved in water. In solution the slightly apolar cyclodextrin cavity is occupied by water molecules, which is energetically unfavourable (polar–apolar interactions) and they are therefore readily substituted by appropriate 'guest molecules' which are less polar than water (Szejtli, 1969; Bergeron *et al.*, 1977). The inclusion complexes thus formed are relatively stable, their water solubility compared to pure β-cyclodextrin is greatly reduced so they rapidly separate out in crystalline form. The physical and chemical characteristics of the included molecules are significantly modified.

Since covalent bonds do not form between the components, under physiological conditions the complexes are easily dissociated. The simplest method is to stir or shake the aqueous solution (cold or warm, neutral or acidic) of cyclodextrin together with the guest molecule or its solution (Cramer and Henglein, 1956). After equilibrium has been attained water is eliminated by freeze drying, spray drying or by any other convenient method; the microcrystalline product often being separated by filtration.

The hydroxyl groups of the cyclodextrins show different reactivities. The primary hydroxyl groups react most easily whilst the C_3–OH groups are most resistant to substitution. From cyclodextrins crosslinked polymers can be prepared in the form of fibrous (Szejtli *et al.*, 1978), block (Solms and Egli, 1965), film (Wiedenhof, 1969) and bead (Wiedenhof *et al.*, 1969) polymers. Such polymers make a variety of specific inclusion chromatographic separations possible.

In the pharmaceutical industry cyclodextrin is utilised mainly as an auxiliary substance to improve the stability and absorption qualities of active ingredients in tablets. This makes the production and introduction of numerous active substances possible whose stability, compatibility or absorption features have so far prevented their use. In terms of the pharmaceutical industry the following effects can be achieved by cyclodextrin complexation:

(i) volatile compounds can be stabilised without losses through evaporation,
(ii) liquid compounds can be transformed into crystalline, compressible forms,
(iii) solubility in water as well as the rate of dissolution of poorly soluble substances can be increased,
(iv) bad taste and smell can be masked by complex formation (Hamilton and Heady, 1970),
(v) cyclodextrin inclusion complexation offers almost full protection against atmospheric oxidation of easily oxidisable substances.

The application/potential application of cyclodextrin inclusion complexes in the food industry can be essentially divided into two categories. First, diminished storage losses (inhibition of volatility and decomposition induced by oxygen, light or heat) and secondly the elimination of disagreeable tastes and smells or other noxious components.

Essential oils extracted from well known spices (dill, tarragon, marjoram etc.) form stable crystalline complexes with cyclodextrins. These aromatics generally consist of a large group of components but by adequate technology an unaltered composition can be assured allowing the same taste sensation to be attained as with the corresponding fresh spice. The cyclodextrin complexes of volatile, readily oxidisable aroma-substances (flavourings) are stable to oxygen, light and heat, however, in aqueous media, e.g. in the mouth, they dissociate instantly and their flavouring action is readily experienced. Complexes of essential oils are utilisable in cosmetics, dairy products etc. More information on cyclodextrins can be found by reference to Szejtli (1988) and Nakamura and Hara (1993).

3.9 Conclusions

As our knowledge of starch structures increases, our understanding and ability to tailor macromolecular compounds and mixtures to the requirements of end users also prospers. Maltodextrins have found uses in a plethora of diverse applications which require slightly differing physical (and sometimes chemical) properties. Their uses will undoubtedly continue to expand as more alternative technologies and methodologies using relatively cheap and abundant starting materials are explored, particularly in areas not normally associated with carbohydrates usage.

References

Altschul, A.M. (1989) *Food Technol.* **43**, 4.

Ammeraal, R.N., Delgado, G.A., Tenbarge, F.L. and Friedman, R.B. (1991) *Carbohydr. Res.* **215**, 179.

Anon. (1988), Paselli SA2, Avebe Product Information, Ref. No. 05.12.31.167EF, Veendam, Holland.

Anon. (1989a) *Maltodextrin 184.1444*, Code of Federal Regulations, 21, parts 170–199, p. 456, Office of the Federal Register National Archives and Records Administration.

Anon. (1989b) Maltrin® M040, Maltodextrin Product Data Ref. No. 0212387, Grain Processing Corp., Muscatine, IA.

Anon. (1990a) *Calorie Control Comment*, Calorie Control Council, **12**, 1, 5.

Anon. (1990b) *Food Technol.* **44**, 3, 92.

Anon. (1991) *Prep. Foods* **160**, 9, 127.

Anon. (undated (a)) *Maltrin® Maltodextrins and Corn Syrup Solids*, Bulletin 11005, Grain Processing Corp., Muscatine, IA.

Anon. (undated (b)) Maltodextrins-Corn Syrup Solids, Bulletin 10003, Grain Processing Corp., Muscatine, I.A.

Anon. (undated (c)) N-Oil® II, Technical Service Bulletin 15889–238, National Starch, Chemical Division, Bridgewater, N.J.

Armbruster, F.C. and Harjes, C.F. (1969) *Chem. Abstr.* **70**, 48867K.
Armbruster, F.C. and Harjes, C.F. (1971) US Patent 3 560 343.
Barker, S.A., Kennedy, J.F. and Somers, P.J. (1968) *Carbohydr. Res.* **8**, 482.
Ben-Bassat, A.I. and Grushka, E. (1991) *J. Liq. Chromatogr.* **14**, 1051.
Bergeron, R.J., Channing, M.A., Gibeily, G.J. and Pillor, D.M. (1977) *J. Am. Chem. Soc.* **99**, 5146.
Chaplin, M.F. and Bucke, C. (1990) *Enzyme Technology*, Cambridge University Press, Cambridge, p. 72.
Cramer, F. and Henglein, M. (1956) *Angew. Chem.* **68**, 649.
Derler, H., Hörmeyer, H.F. and Bonn, G. (1988) *J. Chromatogr. Anal. Biochem.* **440**, 281.
Folkes, D.J. and Taylor P.W. (1982) In *HPLC in Food Analysis* (ed. R. Macrae), Academic Press, London, p. 149.
French, D. (1957) *Adv. Carbohydr. Chem.* **12**, 189.
Hamilton, R.M. and Heady, R.E. (1970) US Patent, 3 528 819.
Hernandez, L.M., Ballou, L. and Ballou, C.L. (1990) *Carbohydr. Res.* **203**, 1.
Heyraud, A. and Rinaudo, M. (1991) *Biotechnology of Amylodextrin Oligosaccharides* (ed. R.B. Friedman), ACS Symposium Series 458, p. 171.
Hunt B.J. and Holding S.R. (1989) *Size Exclusion Chromatography*, Blackie Academic, Glasgow, p. 3.
Kaper, F.S. and Gruppen, H. (1987) *Food Technol.* **41**, 3, 112.
Kennedy, J.F. and White, C.A. (1983) *Bioactive Carbohydrates in Chemistry, Biochemistry and Biology*, Ellis Horwood, Chichester, p. 76.
Klis, J.B. (1984) *Food Technol.* **38**, 11, 92.
Koizumi, K. Kubota, Y., Tanimoto, T. and Okada, Y. (1989) *J. Chromatogr.* **464**, 365.
LaBell, F. and Duxbury, D. (1990) *Food Processing* **51**, 8, 88.
Lim, C.K. (ed.) (1986) *HPLC of Small Molecules*, IRL Press, Oxford, p. 1.
Macrae, R. (ed.) (1982) *HPLC in Food Analysis*, Academic Press, London.
Morehouse, A.L., Malzahn, R.C. and Day, J.T. (1972a) *U.S. Patent* 3 663 369.
Morehouse, A.L., Malzahn, R.C. and Day, J.T. (1972b) *Chem. Abstr.* **77**, 33147W.
Morris, C.E. (1984) *Food Eng.* **56**, 7, 48.
Nakamura, N. and Hara, K. (1993) *Oligosaccharides – Production, Properties and Applications*, Japanese Technology Reviews, Section E: Biotechnology (ed. T. Nakakuki), Gordon and Breach Science, p. 25.
Pszczola, D.E. (1991) *Food Technol.* **45**, 8, 262.
Schierbaum, F.R., Vorwerg, W., Kettlitz, B. and Reuther, F. (1986) *Nahrung/Food* **30**, 1047.
Schwarzenbach, R. (1976) *J. Chromatogr.* **117**, 206.
Scott, F.W. and Hatina, G. (1988) *J. Food Sci.* **53**, 264.
Simpson, J.M. (1993) *Protein Purification Process Engineering* (ed. R.G. Harrison), Marcel Dekker, p. 209.
Solms, J. and Egli, R.H. (1965) *Helv. Chim. Acta* **48**, 1225.
Szejtli, J. (1969) *Die Stärke* **21**, 163.
Szejtli, J. (1988) In *Cyclodextrin Technology*, Kluwer.
Szejtli, J. and Budai, Z. (1976) *Acta Chim. Acad. Sci. Hung.* **91**, 73.
Szejtli, J., Fenyvesi, E. and Zsadon, B. (1978) *Die Stärke* **30**, 127.
Szepesi, G. (1992) *How to use Reverse-Phase HPLC*, VCH, Weinheim, p. 19.
Vorwerg, W., Schierbaum, F.R., Reuther, F. and Kettlitz, B. (1988) In *Biological and Synthetic Polymer Networks* (ed. O. Kramer), Elsevier Applied Science, London, p. 127.
Wang, W., Lundgren, T., Lindh, F., Nilsson, B., Grönberg, G., Brown, J.P., Mentzer-Dibert, H. and Zopf, D. (1992) *Arch. Biochem. Biophys.* **292**, 433.
Wells, G.B. and Lester, R.L. (1979) *Anal. Biochem.* **97**, 184.
Wiedenhof, N. (1969) *Die Stärke* **21**, 163.
Wiedenhof, N., Lammers, J.N.J.J., Van Panthaleon, C.L. and Van Eck, B. (1969) *Die Stärke* **21**, 119.
Ziesenitz, S.C. and Siebert, G. (1987) In *Developments in Sweeteners – 3* (ed. T.H. Granby), Elsevier Applied Science, New York, p. 109.
Zsadon, B., Szilasi, M., Szejtli, J., Seres, G. and Tüdös, F. (1978) *Die Stärke* **30**, 276.
Zsadon, B. Otta, K., Tüdös, F. and Szejtli, J. (1979) *J. Chromatogr.* **172**, 490.

4 Analysis of glucose syrups

J.A. WILSON, G.A. MITCHELL, D.S.J. SPRUYT
and M.J.P. VANDERBIST

4.1 Introduction

Glucose syrups and their related products (e.g. dextrose, maltodextrins) are composed of a theoretically infinite mixture of saccharides based on D-glucose (dextrose), together on occasions with fructose. In the case of sugar alcohols, these saccharides are hydrogenated. As well as the major carbohydrate constituents, there are trace contents of inorganic salts and nitrogenous compounds, and other trace organic compounds, e.g. acids.

As well as the chemical constituents, and in many cases related to them, there is a range of physical properties which are measured, to both monitor and control the production process. These properties are also important in the application of these starch hydrolysis products in formulated foodstuffs.

Particularly in the case of monitoring and controlling production processes, but also useful for other applications, the availability of rapid and reliable methods of analysis is desirable. Developments in the last decade have essentially been in methods of analysis which are more rapid than those which have been replaced.

This chapter is divided into two main sections. The first section describes mainly the measurement of the physical properties and minor constituents of starch hydrolysis products. Many of these methods have been in use for a number of decades, although there are also some where recent developments have led to more rapid results being attained. This section also includes the use of enzymatic methods for the determination of individual sugars. The other major section is devoted to the determination of the saccharide constituents of the products, largely by the application of chromatographic methods.

4.2 Traditional physical and chemical analyses

4.2.1 Dry substance

The determination of dry substance is an important property of a glucose syrup from the point of view of both the producer and the user of the syrup. The methods which are, or have been, used are many and various, but those covered here are the methods most commonly in use today.

4.2.1.1 Refractive index. The concentration of a pure substance dissolved in water is proportional to its refractive index when measured at a specific temperature. In the sugar industry this relationship has been used for many years to measure the solids content of sugar solutions, and refractometers are frequently calibrated with both refractive index and sucrose solids (° Brix).

Glucose syrups are not solutions of single substances and this term covers starch hydrolysis products with a wide range of constituent saccharide contents. Frequently inorganic ions from both the original starch source and those introduced during processing are also present. It is therefore necessary to know the composition of the syrup before the refractive index can be used to determine the dry substance of a syrup.

Notwithstanding this apparent drawback, refractive index is the most commonly used method in the industry for determining the solids content of glucose syrups of known composition. This use is due to its speed and ease of measurement. Sets of tables have been published for converting measured refractive index to dry solids content for a wide range of saccharide compositions, including fructose, and also with ash corrections (CRA, 1975; Wartman *et al.*, 1976; ISO, 1982). The utility and accuracy of the method has also been improved with the derivation of an equation, which is available in a computer software program. This produces a refractive index/solids table for any combination of sugars and trace inorganics content (CRA, 1993).

Measurement is made using a refractometer calibrated with indices from 1.3 to at least 1.6 and accurate to 0.0001 unit. The prism is usually maintained at 45°C ± 0.5°C with circulating water, in order to prevent crystallisation of high dextrose syrups and ease the handling of viscous, high solids syrups. Expression of the result at 20°C, obtained from tables, is also frequently carried out. In practice, a small amount of the sample is introduced onto the prism face of the instrument, which is then immediately closed to prevent evaporation. On achieving temperature equilibrium the reading is noted and the dry substance obtained from the appropriate table.

4.2.1.2 Oven drying. Results obtained from loss in weight on drying will be affected by one or both of the following factors:

(i) Extent of loss of water and other volatiles from the sample.
(ii) Production of water and other volatiles by chemical changes induced by the condition of drying. This, in theory at least, can partially be offset by an increase in weight due to reactions between the substance and oxygen or other environmental substances.

Oven drying methods must therefore be a compromise of the conditions such that the release of water vapour is maximised and the occurrence of chemical reactions is minimised. The method given below is a recognised standard (ISO, 1980; CRA, 1989a), but it should be stressed that in order to achieve reproducible results it is essential that details of the method are obtained from the original source and are strictly adhered to. There is also a degree of controversy about the applicability of oven drying methods to glucose syrups, and for that reason it is nowadays not commonly used.

The method utilises quartz sand or diatomaceous earth as an adsorbent for the diluted sample. These are mixed together in a lidded aluminium moisture dish with a small glass rod and all are weighed. This is dried to constant weight at 100°C in a vacuum oven, the sample being stirred with the glass rod to break up any lumps after about 5 h, before being placed back in the oven.

4.2.1.3 Moisture by Karl Fischer titration. This method is based on chemical reactions which are reasonably specific for water, although some other compounds may interfere. In principle the method depends upon the water in the sample being taken up by a solvent containing methanol and formamide, the excess then being reacted with iodine in the titrant (CRA, 1992; ISO, 1983; MacLeod, 1991).

It is usual for commercially available instrumentation to be used nowadays, together with modified reagents which obviate the need for the noxious pyridine originally used. The end point may be detected visually, but it is more usual to determine it electronically. The reagents are standardised against a standard solution of water in methanol, or a pure salt hydrate such as sodium tartrate dihydrate may be used.

The method has the advantage of producing a direct measure of water content and when in routine use is a relatively rapid method. It does, however, have drawbacks in that, to obtain accurate and consistent results, great care must be exercised in order to prevent the ingress of extraneous moisture at any stage.

4.2.1.4 Specific gravity/density. The use of hydrometers calibrated to the Baumé scale was a common method, originating in the USA, of indirectly determining the solids content of glucose syrups. If the DE of the syrup is known, tables are available for converting the Baumé to solids (CRA, 1959). In order to overcome the effects of high viscosity and entrapped air, the specific gravity is determined at elevated temperature (usually 140°F) to ensure consistent results. An arbitrary 1.0° Baumé correction factor is added to reduce the value to that which would be obtained at 100°F. The method is rarely used nowadays other than for manufacturing process control, and rather than hydrometers, in-line density meters are more generally employed (e.g. utilising the principle of absorption of ionised radiation).

4.2.2 *Dextrose equivalent (DE)*

The definition of dextrose equivalent is 'the reducing sugar content expressed as dextrose percentage on dry matter', using a standard method for the determination of reducing sugars. By this definition pure starch has zero DE, whilst dextrose (D-glucose) has a DE of 100. The DE of a starch hydrolysate is therefore an indication of the degree of hydrolysis of the starch, although with the advent of enzymatically hydrolysed syrups containing sugars other than dextrose as the predominant sugar, the relationship is no longer as meaningful, as syrups of very different composition can have the same DE.

4.2.2.1 Lane–Eynon titration. The standard method for the determination of DE relies on the reducing power of the carbohydrate, which reduces copper sulphate in an alkaline tartrate solution. The method used is the constant volume Lane and Eynon method (CRA, 1968), which will not be detailed here. Suffice it to say that accurate and reproducible results are only obtained by strict adherence to the detail of the method, such as boiling time, total elapsed time and volume of sample titrant. In experienced hands and with samples of known approximate DE it is a rapid and accurate method.

4.2.2.2 Osmometry. The determination of DE or degree of hydrolysis is, in effect, a determination of the average number molecular weight of a starch hydrolysate. The colligative properties are therefore directly related to DE, and this fact is made use of by the determination of DE from the depression of freezing point, using an osmometer (Kearsley, 1978). This is a method which has become increasingly commonly utilised for process control, due to its speed, although a correction for ash content has to be applied.

The depression of freezing point of a solution of the saccharide sample is accurately determined by use of a programmable, commercial osmometer, and the DE is read off from a calibration graph prepared using samples of known DE by Lane–Eynon titration (Fitton, 1979).

4.2.3 *Sugars determination by enzymic methods*

The specificity of enzyme induced reactions provides a very interesting extra dimension for the measurement of sugars. Glucose, in particular, lends itself readily to enzyme based procedures. There are two approaches which form the basis of this analytical concept. One protocol utilises the action of glucose oxidase (GOD) on glucose to produce gluconic acid and hydrogen peroxide. Only β-D-glucose is involved, so it is imperative that the sample is at the equilibrium point (67:33, β:α) for glucose in solution. The hydrogen peroxide produced, in the presence of peroxidase, will then induce a change in the chromophore of a coloured substance which is measured spectrophotometrically. A suitable reagent for this is 2,2-azino-di-[3-ethylbenzthiazoline sulphonate (6)], commonly known as ABTS, which produces a green colour (AFNOR, 1986). Another suitable reagent is 4-aminoantipyrine (Megazyme, 1992) which yields a red colour.

An original approach has also been developed, by using a cell containing a membrane with immobilised glucose oxidase (Y.S.I., 1979). When the sample is injected, glucose passes through the membrane where it reacts with GOD to produce hydrogen peroxide. This, in turn, traverses to a platinum anode where it is oxidised. The current flow produced is linearly proportional to the local concentration of hydrogen peroxide at the equilibrium point of the reaction and this makes the system a very fast means of quantifying glucose. In the industry this is used for process monitoring; correlation against HPLC is very good. This approach is limited by the availability of different oxidases and as yet no suitable method exists for fructose.

A second protocol uses the reduction of nicotinamide adenine dinucleotide phosphate (NADP) as the final measurement step. The change in concentration of NADPH can be measured spectrophotometrically in the UV waveband. Both glucose and fructose undergo phosphorylation by adenosine-5-triphosphate (ATP) when catalysed with hexokinase (HK). The glucose-6-phosphate (G-6-P) formed is oxidised by NADP with the formation of NADPH which is stoichiometric with the amount of D-glucose. To determine fructose the fructose-6-phosphate (F-6-P) formed in the first step must be isomerised to G-6-P, and then the same determination made as for glucose. This explanation is simplified and the concept has been fully exploited by Boehringer, who sell kits for several sugars. Many of the procedures rely on transformation of the original sugar (e.g. maltose, saccharose) to yield glucose initially.

Both enzymic approaches have advantages and disadvantages. Using a colorimetric measurement for the glucose requires calibration and certain steps of the total analysis may not necessarily be specific. There is no doubt about the specificity of the HK approach. All methods are extremely sensitive, which may mean working with very dilute solutions and small volumes. This is especially so for the Boehringer protocol. All the reactions take place in the cuvette of the spectrophotometer and often sample size is 0.1 ml. Our own experience shows that if this aliquot can be increased to the maximum possible (2 ml), repeatability is greatly improved. None of the commercial kits appears to have been properly validated with regard to tolerance of the method. The enzymic methods should be viewed as complementing chromatographic methods.

4.2.4 Acidity

There is a small residual acidity in glucose syrups, due to trace amounts of organic acids and buffer salts produced from them with inorganic cations. With ion exchanged glucose syrups this acidity is of little significance, but it is still determined for a few applications. There is a standard method agreed within the industry (CRA, 1983), whereby a 40% w/w solution of the sample is titrated with 0.05 M sodium hydroxide solution to a pH 6.0 end point. The result is expressed in meq g^{-1}.

4.2.5 Ash

Ash present in glucose syrups originates from both the starch source and also from any acid treatment of the starch, or pH adjustments, during hydrolysis and refining. As mentioned earlier, it is now common practice to ion exchange refined glucose syrups, so residual ash contents are normally very low.

It is usual to determine sulphated ash in glucose syrup. This aids in the decomposition of the organic matter in the early stages of the ashing process and leads to a more consistent ash. Standard methods of sulphated ash determination are used (CRA, 1985; ISO, 1994), the initial charring with acid taking place in a silica dish on a hotplate or over a Bunsen flame in a fume cupboard. Care has to be taken initially to prevent loss of sample due to foaming. The dish is then placed in a muffle furnace at 550°C until constant weight is achieved. Sulphated ash is sometimes converted to 'true' ash by multiplying by the factor 0.85.

4.2.6 Inorganic ions

As ion exchanged glucose syrups contain such low contents of inorganic ions it is now generally not necessary to determine any ions in a frequent, routine manner, although there are still a few applications where this may be needed. The individual ions determined by chemical means, in this context, are calcium and chloride, but the advent of ion exchange chromatography has enabled the determination of a wider range of both anions and cations at very low levels. This latter subject is covered in section 4.3.

4.2.6.1 Calcium. The level of calcium in glucose syrups is critical in some applications, e.g. in a positive context in systems where the gelling of hydrocolloids is determined by the content of calcium ions, and in a negative context where calcium ions may be associated with the occurrence of haze development. The standard method used (CRA, 1973) consists of ashing the sample with sulphuric acid to destroy the organic matter. The residue is dissolved in dilute acid, and potassium hydroxide and triethanolamine are added to prevent interference by iron, magnesium and other cations. The solution is titrated with ethylene glycol tetraacetic acid (EGTA), using hydroxynaphthyl blue indicator for determination of the end point. Titration with EDTA is also possible.

4.2.6.2 Chlorides. The chloride content of a sample is determined by titration with a standard silver nitrate solution, using potassium chromate as an indicator (CRA, 1990a). Alternatively the end point may be detected by the use of a silver–silver chloride electrode.

4.2.6.3 Ion chromatography. This topic is dealt with in more detail in section 4.3.4.8, where its application in relation to the determination of individual saccharides and oligosaccharides is covered. The technique was developed first however for the determination of inorganic ions, and the Dionex™ system in particular is widely used for this purpose.

Ion chromatography columns are packed with resin beads with a pellicular ion-exchange surface coating. These coatings carry a tertiary amine function in the case of anion resins and a sulphonic acid function in the cation resins. The eluents generally used for ion exchange chromatography are sodium hydroxide (typically 1–100 mM) and methanesulphonic acid (typically 20 mM) respectively for anion and cation analysis. Isocratic elution is most commonly used, but gradient elution can be used to decrease the retention time of more strongly retained ions.

Suppression is used in order to reduce the eluent background and increase the conductance of the sample components, by conversion to the acid or hydroxide forms. Suppressors consists of a derivatised microporous membrane. These membranes have exchange sites for ions having a charge opposite to that of the analyte.

Suppression comes in two forms, chemical suppression and autosuppression. Chemical suppression uses a chemical solution to provide the hydrogen and hydroxide ions required, while autosuppression applies a current to the suppressed eluent (or water), causing it to dissociate and provide these ions.

The most commonly used detection system for ion chromatography is the conductivity detector. More sophisticated detectors are available, such as pulsed amperometric detection, but these are more relevant to other analyses using ion chromatography, such as for carbohydrates. This technique is again dealt with in detail in the later section. Test chromatograms of standard anion and cation mixtures are shown in Figures 4.1 and 4.2.

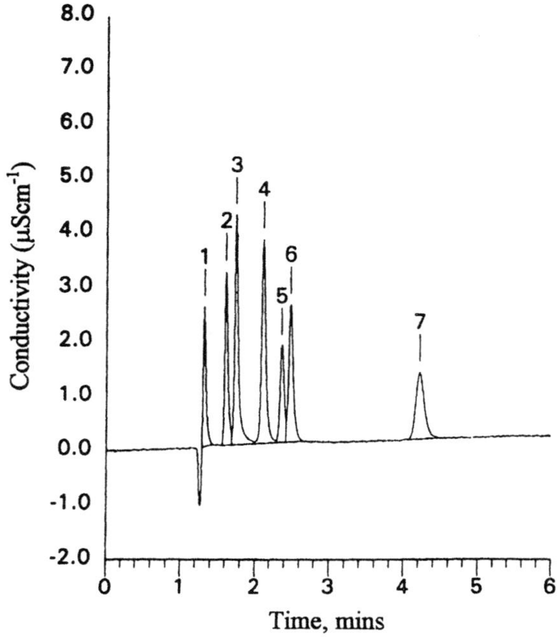

Figure 4.1 Separation of a standard anion mixture on a DIONEX IonPac™ AS11 column. 1, Fluoride; 2, chloride; 3, nitrite; 4, sulphate; 5, bromide; 6, nitrate; 7, phosphate.

4.2.7 Colour

The determinations of colour and colour stability of glucose syrups are critical parameters for both producers and users of the syrups. Reactions between reducing sugars and residual nitrogenous compounds can lead to the formation of colour bodies. The type and efficiency of syrup refining therefore has a major impact on the stability of a syrup to colour development at elevated storage or processing temperatures.

The standard method for the determination of colour (CRA, 1989b) is the spectrophotometric determination of light absorbance at 450 nm. The effect of light scattering due to haze is corrected for by subtracting the absorbance at 600 nm. Due to the low colour absorbance of most glucose syrups, a cell length of at least 4 cm is required, with in some cases 10 cm being preferred, but results are corrected back to a 1 cm cell.

The presence of air bubbles, which in viscous syrups is difficult to avoid, has a negative influence on the accuracy of the method and for this reason the industry frequently uses a visual comparison method. In this method the colour of the sample in a colour comparison tube is visually matched with commercially available standard glasses (e.g. Lovibond

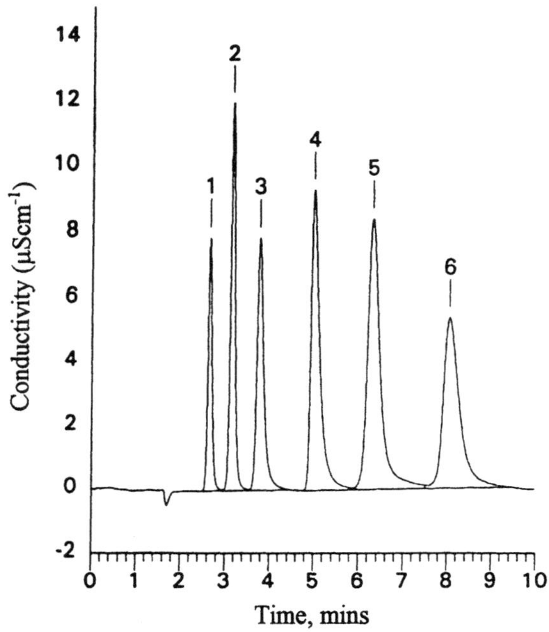

Figure 4.2 Separation of a standard cation mixture on a DIONEX IonPac™ CS12 column. 1, Lithium; 2, sodium; 3, ammonium; 4, potassium; 5, magnesium; 6, calcium.

yellow and red glasses) in a colour comparator. This reading (the sum of the Y and R glasses) is converted to optical density using AOCS tables (CRA, 1990b). In all cases of syrup colour determination, if a diluted sample is used the result is converted to an arbitrary 80% solids by the use of the relevant dilution factor.

Becoming more popular for measuring the colour of syrups are tristimulus instruments. These try to quantify the three colour response mechanisms of the human eye. The tristimulus values obtained are the basis for uniform colour scales (*L*, *a*, *b*), which are based on the opponent colours theory. Experiment shows that a good correlation can be established between Lovibond values and the *b* dimension, which is associated with the yellow to blue response. An advantage of this approach is that the *b* value is little affected by haze or air bubbles.

For determination of colour stability (or heat colour), the sample of syrup at finished solids is maintained at 100°C for a time of between 1 h and 18 h, depending on the syrup type and application. The colour after cooling is then measured as above.

4.2.8 Conductivity and pH

With the general availability of ion exchanged glucose syrups and related products, the determination of pH is more difficult and less meaningful. For this reason the determination of conductivity is often more meaningful. There is not a finalised standard method within the industry, so concentration and temperature may vary, both of which have a significant effect on the values obtained. Values of 30% solids and 25°C are fairly common however for these two parameters. Conductivity is determined using a commercial conductivity meter, results being expressed in $\mu S\ cm^{-1}$.

The determination of pH is usually carried out on a 50% w/w solution of the syrup. For samples of low inorganic ion content it is necessary to leave the pH electrode undisturbed in the sample for up to 5 min to allow a stable reading to be achieved. It is also important to ensure that there is good electrical continuity between the sample and the reference electrode (e.g. the 'frit' or other junction between the sample and the saturated potassium chloride in the reference electrode can become blocked and lead to false results).

4.2.9 Sulphur dioxide

Glucose syrups produced from maize starch obtained by the wet milling process will contain trace quantities of sulphur dioxide, residual from the steeping process in dilute sulphurous acid solution. Sulphur dioxide is also frequently added to glucose syrups at the final evaporation stage, not primarily to prevent or minimise the growth of microorganisms, but to limit the colour development in storage and/or in the processing of the glucose syrup ingredient into the finished foodstuff. A typical example of this is the cooking of sugar (sucrose) and glucose to produce high-boiled sweets. The sulphur dioxide is usually added to the glucose in the form of sodium metabisulphite solution.

The standard method for sulphur dioxide determination is the modified Monier Williams method (CRA, 1991). In this method sulphur dioxide is removed from the acidified, boiling sample in a stream of nitrogen. The gas stream is then passed through a dilute hydrogen peroxide solution which oxidises and traps the sulphur dioxide to produce sulphuric acid. This is titrated against a standard sodium hydroxide solution. Alternatively a gravimetric analysis may be carried out after reacting the sulphuric acid with barium chloride to produce the insoluble sulphate. This latter addition, whilst being very time consuming, can overcome the problems of interfering organic acids and related substances. An example where this can be a problem is in the case of high fructose-containing samples. Fructose in acidic solution at elevated temperatures is very prone to be converted to 5-hydroxymethylfurfural (HMF), which in turn yields formic acid. Humines (coloured caramel-like products) are also formed, which decompose to volatile acidic substances. These acids will lead to falsely high SO_2 results being obtained by titration with alkali.

An alternative more rapid method for the determination of sulphur dioxide is the direct titration in an acidified solution with a standard iodine solution (CRA, 1988). Bound sulphur dioxide is first released by treatment of the sample with alkali in an ice chilled bath. After the subsequent acidification, the titration must be completed rapidly in order to minimise the recombination of the sulphur dioxide.

4.2.10 Nitrogen and protein

Protein determinations play an important part in the process control of most starch separation processes, and the quality control of animal feed coproducts. Small residual amounts of nitrogenous compounds also play a critical role in the development of colour in glucose syrups and the monitoring of organic nitrogen provides a useful control check. Kjeldahl nitrogen is used for this purpose, but its relatively poor sensitivity does not fit well with the very low levels found in glucose syrups. The use of a total nitrogen analyser is more suitable for this application, and the principles of this will also be described. Both these methods yield results for nitrogen content and this is usually converted to protein by multiplying by the factor 6.25 for maize and potato products, and by 5.7 for wheat-derived products.

4.2.10.1 Kjeldahl nitrogen. Samples are first digested in boiling sulphuric acid containing potassium sulphate to raise the boiling point and a catalyst (usually copper selenite). The proteinaceous nitrogen is converted to ammonium sulphate. The digest, after cooling and dilution, is made alkaline and distilled into a standard acid solution. The residual acid is then back titrated with standard alkali, using methyl red indicator. Alternatively the ammonia can be distilled into a boric acid solution, followed by titration with acid (CRA, 1980). The usual indicator for this method is methyl red/bromocresol green mixed indicator, but the titration is dilution sensitive, and distillate volumes of samples and blanks should be adjusted to constant volume with distilled water prior to titration. This titration is the basis of a number of commercial protein analysers.

4.2.10.2 Total nitrogen analyser. One method used in the design of total nitrogen analysers is based on the catalytic oxidative pyrolysis of the sample and chemiluminescence for detection, and it is sensitive to levels down to approximately 0.01 mg kg^{-1}. The sample is injected into a reaction furnace containing a platinum catalyst along with a stream of oxygen. The furnace is set to a temperature in the range of 800–900°C, depending on the type of sample. Under these conditions nitrogen compounds are pyrolysed to nitric oxide. This gas stream is mixed with a stream of ozone from an ozone generator, under which conditions the following chemical reaction occurs:

$$NO + O_3 \rightarrow NO_2 + O_2 + h\nu$$

The luminescence is detected and the signal handled by a data processing unit. The instrument is calibrated with a standard ammonium sulphate solution when being used for aqueous samples such as glucose syrups. An example of a total nitrogen analyser suitable for glucose syrups is the TN-05 produced by the Mitsubishi Kasei Corporation. An alternative detection methodology used in other commercial instruments depends on the reduction of nitric oxide to nitrogen, over a copper catalyst. The content of nitrogen is then determined by a thermal conductivity detector.

4.3 Determinations using chromatographic methods

4.3.1 Introduction

A knowledge of the saccharide distribution in glucose syrups is of prime importance, due to the ability to 'tailor-make' glucose syrups. Whereas acid used to be the sole hydrolysis medium for starch, a host of enzymes is now commercially available. These provide the manufacturer with a wide choice of processes. In addition, the customer has become more informed and will often demand quite strict specifications for the chromatographic profile of a product.

For some uses such as chemical synthesis, the ultimate level of dextrose in a syrup is of most importance. At the same time, the nature of the residual polysaccharide is relevant, as it could cause haze during the customer's process. Reversion of dextrose can occur during processing and this is monitored by determination of the levels of isomaltose and maltulose. High maltose syrups are now popular adjuncts in the brewing industry, and accurate determination of dextrose, maltose and maltotriose is essential. The fructose content in isomerised glucose syrups must be quantified precisely to meet legal requirements concerning levels in the end product and during processing.

The viscosity behaviour of a syrup during a customer process must be within certain bounds. To gauge this parameter, the molecular weight distribution (MWD) provides a unique indicator. Health awareness has promoted the production of hydrogenated sugars and syrups as sweeteners because the human digestive system metabolises them only slowly, if at all. In addition, the range of sugar alcohols commercially available has increased. Changes in agricultural politics have made US maize an expensive commodity in Europe, and this in turn has favoured the use of wheat for starch production. Different starch types often require modified processes to provide acceptable syrups. There may be subtle differences between syrups from different starch types.

The analyst must cope with a broad range of needs as demanded by the raw material, the process, the end product and the customer requirements. How this is achieved is described in the subsequent sections.

4.3.2 Overview of chromatographic methods

Since 1980 at least, there has been almost universal adoption of high performance liquid chromatography (HPLC) for the analysis of glucose syrups. The advantages so greatly outweigh any disadvantages that other techniques such as gas liquid chromatography (GLC) are employed only on rare occasions. Within HPLC there has been a rapid evolution of columns, detectors, and methods, which has improved its suitability in almost every sense.

The resolving power of GLC is still superior to that of HPLC for sugar determination. High oligomers and polysaccharides are better quantified by HPLC. When capillary columns are used on GLC an even higher resolution is obtained. However, the necessary steps involved in gas chromatography such as derivatisation, calibration, and column handling are decided drawbacks. The detection, with flame ionisation (FID) is very sensitive, but often a sugar elutes as several anomers, making quantification of the parent sugar difficult. Most of the problems can be handled by, for example, including an internal standard, carefully monitoring the quantities of sample and choosing a silylating reagent that inhibits the number of derivatives. Packed columns are invariably based on low polarity silicones. Folkes and Brookes (1984) gave a full account of the methodology for sugar analysis with GLC.

The application of HPLC progressed when modified silica columns, and in particular amino-bonded silica, became available. Acetonitrile/water is the solvent, and by varying the ratio, a resolution is achieved which compares with GLC. Sensitivity of the system is still lower, due to the use of the differential refractometer. Columns packed with ion exchange resins were already under investigation by Brobst *et al.* (1973) and their use brought two outstanding advantages. These were that water could be used as eluent and that the whole sample matrix eluted from the column. Work on improving the resins continued, and Fitt *et al.* (1980) could separate up to the pentasaccharides on a 4% crosslinked polystyrene resin, sulphonated and in the calcium form. Speed of analysis was also becoming faster. The resin producers optimised the physical nature and chemical composition of the resins and custom packed columns became available. Initially calcium was the sole counterion but investigation revealed that other cations conferred different resolving power to the resin. Now there are many forms of cation exchange resins being used, including H^+, Na^+, K^+, Ag^+, Ca^{2+}, Zn^{2+} and Pb^{2+}. These provide a powerful inventory of columns to satisfy the earlier mentioned analytical needs. Also, detectors have become more sensitive.

A later introduction to the field is the so named 'ion chromatograph'. Originally developed for ion analyses using anion columns and conductivity detection (as mentioned in section 4.2), it was expanded to cope with sugar analyses by the development of the electrochemical detector. The resolution of this technique begins to approach that of GLC. Obviously HPLC is established as the major method for analysis of sugars and glucose syrups. The following sections will describe the present status of GLC and will give a full description of the current application of HPLC in the glucose syrup industry.

4.3.3 Gas chromatography of carbohydrates

4.3.3.1 Introduction. Gas chromatography was already well established before carbohydrates could be separated in a reasonable way by this technique. A major prerequisite for analysing substances by gas chromatography is that the analyte has to be sufficiently volatile to be brought into the gas phase. This was certainly not the case for carbohydrates before adequate derivatisation was developed to make them volatile. Before this development, the lengthy and tedious technique of paper chromatography was used.

As methods to derivatise sugars became known, gas chromatography for the analysis of carbohydrates became a widely used tool. However, by the introduction of HPLC, the use of gas chromatography declined slowly for sugar analyses. Gas chromatography is only used for the analysis of carbohydrates as a matter of convenience, or where HPLC is inadequate. This does not imply that no interesting developments have occurred in the last decade. Inherent in the general development of gas chromatography, the appearance and increased use of capillary columns was already a major step forward. In addition to that, more recent developments such as high temperature gas chromatography (HT-GC), made the basic technique again interesting for carbohydrate analysis. For structure analysis of carbohydrates, GC still remains a valuable tool. A review, including other techniques, has been made by Robards and Whitelaw (1986).

4.3.3.2 Derivatisation. The starting point in GC analysis of carbohydrates is the derivatisation. A breakthrough in solving the problem of volatility and thermal stability of derivatives was first studied by Sweeley *et al.* (1963). The development, involving many researchers, finally resulted in a generally adopted methodology focusing mostly on the use of trimethylsilyl (TMS) derivatives. Even earlier work by Bourne *et al.* (1950) proposed other derivatives and these still find application.

The base reagent to obtain the derivatives was for a long time hexamethyldisilazane (HMDS), but nowadays the use of *N*,*O*-bis(trimethylsilyl) trifluoroacetamide (BSTFA) is probably the most widespread derivatisation reagent for GC of carbohydrates. Many other similar reagents and mixtures have been developed and tested for different uses, depending on the type of carbohydrate separation needed. A major factor to keep in mind is which kind of anomer separation is wanted or should be avoided. Literature and suppliers provide the best information for the choice. For sugar alcohols, the simplest approach is the esterification to acetates by means of acetic anhydride under reflux. Permethylated carbohydrate derivatives are mainly used in structure analysis of complex carbohydrates.

4.3.3.3 Columns. Perhaps the era of packed columns is finished, but often as a matter of convenience their use is still valid. In packed or capillary columns the liquid phase is often of the same nature. For the separation of TMS derivatives, non-polar liquid phases are used. The main types are polydimethylsiloxane (SE 30, OV-1, OV-101) or polyphenylmethylsiloxane (OV-17), and polycyanopropylphenylmethylsiloxane (OV-1701).

When analysing acetate derivatives, more polar liquid phases are used; typical are polyethyleneglycol phases (Carbowax types and similar). The breakthrough in the use of capillary columns came with fused silica open tubular columns reinforced with a polyimide coating on the outside. This novelty removed the fragility and other drawbacks of the glass capillary columns. Chemically bound liquid phases on capillary columns are now commonly used and have better baseline stability at high temperatures.

Film thickness of the liquid phase varies from 0.25 to 1.5 μm, and even up to 5 μm depending on the application. In general, thicker films show better retention and selectivity and can be an alternative to very long columns. Provided that the instrument oven can reach temperatures over 450°C, HT-GC can be applied. Several developments in this field have been studied by Sandra and David (1989) and Sidisky *et al.* (1988). An advantage of HT-GC is the possibility of separating higher oligomers, up to a degree of polymerisation (DP) of 7 to 8, which was not possible in classical GC techniques.

4.3.3.4 Conclusion. Numerous articles exist on specific applications for the GC analysis of carbohydrates. While novel techniques have redrawn the attention back to GC for certain specific analyses in the carbohydrate field, most analyses of glucose syrups are satisfactorily handled by HPLC.

4.3.4 *Liquid chromatography of carbohydrates*

4.3.4.1 Introduction. As mentioned earlier, HPLC is the predominant technique for the analysis of sugars and starch hydrolysates. There has been development in many facets of the methodology such as columns, detectors and data acquisition, so that sensitivity, selectivity and accuracy, collectively outperform all other techniques. In addition, speed of analysis, ease of sample preparation, use of water as eluent and easy maintenance of the equipment are other bonuses. Besides simple glucose syrup analysis, identification and quantification of a range of sugars is made possible by the diversity of columns, eluents and detectors that can be used. At the other extreme, size exclusion chromatography now allows the analyst to make a detailed fingerprint of the polysaccharide fraction of any hydrolysate.

Two types of HPLC column are used for the separation of sugars, the silica-based columns and the polymer-based columns. The silica-based columns were developed first, but have been slowly superseded by the polymer-based columns.

4.3.4.2 Silica-based columns. The resolution of carbohydrates was achieved at first on silica-based columns. The first silica packing material was irregularly shaped, giving large void volumes. Spherical silica was introduced in the mid-1970s, and smaller and better controlled particle size improved column selectivity and efficiency.

Several systems have been described for the separation of sugars on silica-based columns by Heyraud and Rinaudo (1981), Boumahraz *et al.* (1982), Verhaar *et al.* (1984) and Koizumi *et al.* (1984 and 1987). Most of the methods for carbohydrate determination are based on amino-bonded silica gel columns. The most frequent amino binding is made by derivatising silica gel with aminopropyltriethoxysilane. Separation is achieved using an isocratic elution with an acetonitrile–water eluent. A mixture containing 80% of acetonitrile is commonly used, but the polarity of the solvent can be changed to obtain the desired resolution. Many applications such as gluco-oligomers determination, DP 2 isomers quantification and sucrose analysis have been reported.

Several other amino modified silica systems have been developed, such as the polyamine polymer bonded silica from Yamamura Chemical (YMC-Pack PA Gel) or the carbamoyl modified silica from Toyo Soda (TSK gel Amide-80). A comparison of these two columns with two conventional aminopropyl silica columns has been reported by Koizumi *et al.* (1987).

Another approach uses the *in situ* modification of silica by an amine modifier dissolved in the eluent (Boumahraz *et al.*, 1982; Aitzetmuller, 1978; Wheals and White, 1979). Silica modification with several amine modifiers such as piperazine, ethylenediamine, tetraethylenepentamine

(TEPA) and diaminobutane combined with spermine have been reported. Figure 4.3 illustrates the separation of gluco-oligomers in a starch hydrolysate using straight silica modified with TEPA.

The efficiency of aminopropyl silica columns can decrease rapidly, and this is attributable to the removal of the modifier by the eluent. Adding an amine modifier to the eluent reinforces the selectivity of aminopropyl silica columns. This system is suited to the separation of DP 2 and DP 3 isomers as seen in Figure 4.4. To improve the separation of disaccharides, Muller and Siepe (1980) replaced the acetonitrile–water eluent by a mixture of acetone, ethyl acetate and water.

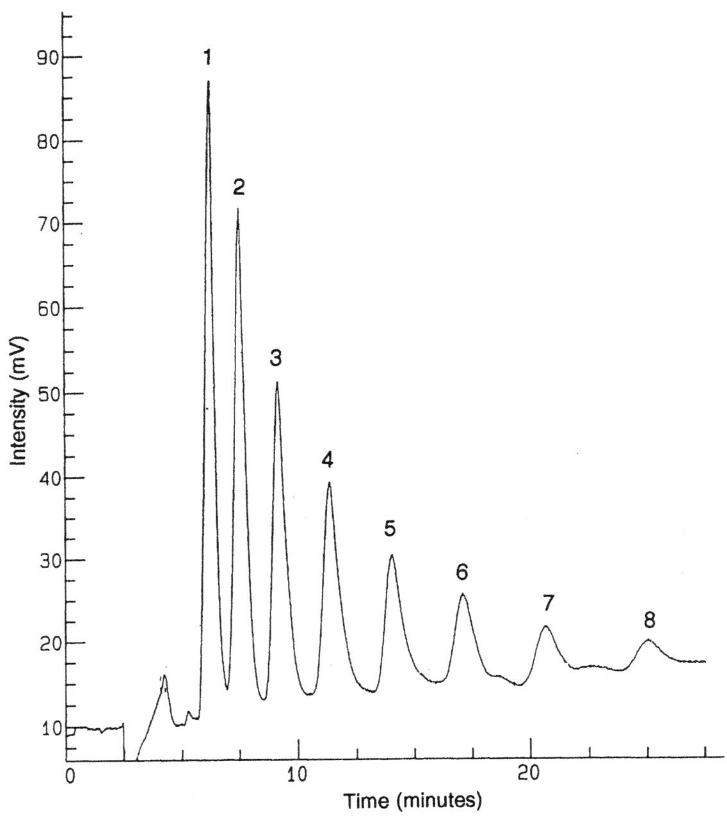

Figure 4.3 Oligosaccharides resolution in acid glucose syrup. Column, Waters Radial-Pak RCM-100, silica 10 μm, 10 cm, room temperature. Solvent, acetonitrile–water 60:40 v/v with 400 mg l^{-1} of tetraethylenepentamine, at 1 ml min^{-1}. Detection, differential refractometer. 1, Dextrose; 2, maltose; 3, maltotriose; 4, maltotetraose; 5, maltopentaose; 6, maltohexaose; 7, maltoheptaose; 8, maltooctaose.

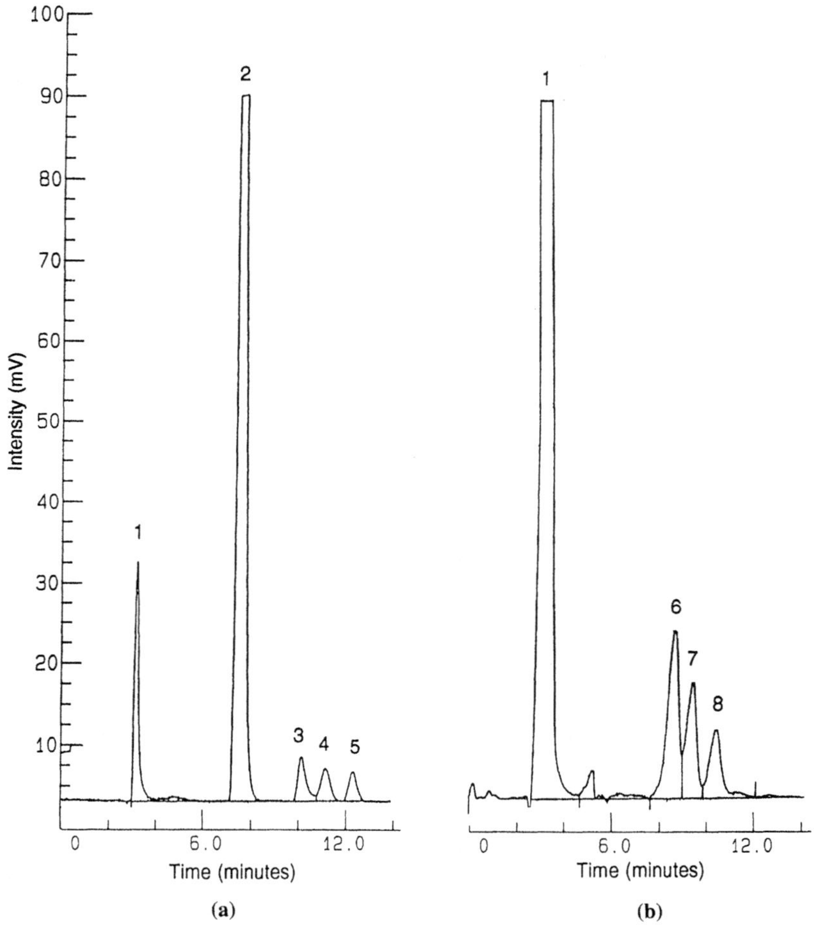

Figure 4.4 DP 2 (a) and DP 3 (b) isomers resolution. (a) Column, Waters Radial-Pak RCM-100, si-NH_2 10 μm, 10 cm × 8 mm, room temperature. Solvent, acetonitrile–water 72:28 v/v with 400 mg l^{-1} of tetraethylenepentamine, at 1.5 ml min^{-1}. Detection, differential refractometer; (b) column, Nucleosil 120-7 NH_2, 10 cm × 4 mm cartridge, room temperature. Solvent, acetonitrile–water 72:28 v/v with 400 mg l^{-1} of tetraethylenepentamine, at 1.0 ml min^{-1}. Detection, differential refractometer. 1, Water; 2, dextrose; 3, maltulose; 4, maltose; 5, isomaltose; 6, maltotriose; 7, panose; 8, isomaltotriose.

Classic reversed phase columns have also been used, and especially octadecyl (C18) silica gel eluted with water (Verhaar *et al.*, 1984). Good separation of mono-, di-, tri- and higher saccharides is achieved. Unfortunately, the resolution between non-glucose monosaccharides is poor, and applications are restricted to the separation of gluco-oligomers.

Here again, anomeric separation can occur, which makes the interpretation of the chromatogram quite difficult. It is possible to overcome this problem by addition of an amine in the eluent. The effect of using different degrees of alkyl substitution and different alkyl chain length has been studied. Finally, ion pairing reagents have been used on reversed phase columns (Lochmuller and Hill, 1983).

Silica-based columns are still used, but apart from the excellent rigidity of the packings, there are several disadvantages. Strong interaction between analyte and packing makes the analysis too long, and high molecular weight materials stay on the column. Acetonitrile is an expensive and somewhat hazardous solvent and removes bonded phases. Silica, in general, dissolves slowly in water, and faster at high pH. At low pH, anomeric separation increases.

4.3.4.3 Polymer-based columns. Polymer-based HPLC packings are today widely used for carbohydrate analysis. Like the silica packing, the first polymer beads were irregularly shaped and column efficiency was poor. One main problem of the first polymer-based HPLC columns was attributable to the low rigidity of the polymer, or to the poor mass transfer. Spherical microparticles with narrow particle size distribution dramatically improved the columns performance.

Custom packed polymer-based HPLC columns for the separation of sugars were introduced in the late 1970s. Conrad and Palmer (1976) reported a rapid analysis using a polymer-based column where the high molecular weight material eluted first and the dextrose last. Nowadays numerous column manufacturers supply a large range of packed columns.

Most of the matrices used in such HPLC columns are gels comprised of a polystyrene–divinylbenzene polymer, with different degrees of cross-linking. These polymers can be sulphonated or aminated to form ion exchange functional groups. They have a very good chemical and mechanical stability for analytical chromatography. Section 4.3.4.7 will deal entirely with these resins, sulphonated, and thus in the cationic form. Anion exchange resins are described and their applications are reported in section 4.3.4.8.

The resins have been optimised with regards to their chemical constitution and particle size. Pore size is controlled by the degree of crossbonding, which determines the exclusion limit of the column. Column dimensions are also important, but perhaps the critical parameter is the counterion used. It is this which has the greatest influence on the interaction between the static phase and the sugars, especially the monomers and the dimers. This observation has been reported by Goulding (1975). Each counterion has its own effect for a specific sugar, and just by changing the counterion, a quite different chromatographic profile for the same syrup can be obtained. The sugars are separated by the ion moderated partition (IMP) chromatography technique, in which several physicochemical phenomena are involved to effect separation. Those mechanisms employed in IMP are ion exclusion, normal and reverse phase partition, size exclusion, and ligand exchange (Bio-Rad, 1994).

4.3.4.4 Sample preparation. Glucose syrups need some preparation before chromatography, and briefly, the sample should be at the right dry substance, free of insoluble material and preferably free of salts.

(a) Dilution. As manufactured glucose syrups have a dry substance of up to 85%, viscosity can be high and crystallisation can occur. Dilutions with water to 5% to 15% dry substance are adequate for HPLC. With modern detectors, the limit of detection is around 0.1 μg of sugar, and for accurate results a peak representing at least 2 μg of sugar is recommended. Such a range allows small quantities to be injected, prolonging the lifetime of the column. Typical injected amounts of syrup are between 100 to 500 μg of dry substance. If crystallisation occurs, the sample must be warmed and totally solubilised before final dilution.

(b) Demineralisation. While commercial syrups become continually more refined, minor components such as protein or ash will still often be present. Salts which are soluble in water will be eluted with the sugars and might interfere in the chromatogram. The cation could affect the ionic form of the cation exchange resin by replacing the counterion fixed onto the sulphonated group. The anion will pass through the column and be detected by the refractometer. Each anion will also elute as its own specific retention time, according to the ionic form of the resin, and can interfere with the sugars. As an example, on a cation exchange resin in the sodium form, chloride elutes at the void volume of the column and interferes with the high molecular weight species. On the same system, phosphate and sulphate elute at the same region of the chromatogram as DP 3 and DP 2, and finally acetate elutes just after the dextrose peak, interfering with fructose, if any. The behaviour of these salts is dependent on the counterion.

Therefore all salts should be removed either by batch ion exchanging or by layered resins in small ion exchange columns. A two-step system can be used, e.g. first removal of cations, then of anions, or a single-step system working with mixed bed resins. The correct contact time between the syrup and the resins has to be determined. Too short times will not remove all salts, but too long times could result in sugars transformation, e.g. isomerisation due to pH changes.

However, the simplest and most convenient method is an in-line deashing system developed by Fitt *et al.* (1980) and manufactured later by Bio-Rad. Two small cartridges, one filled with a cation exchange resin, the

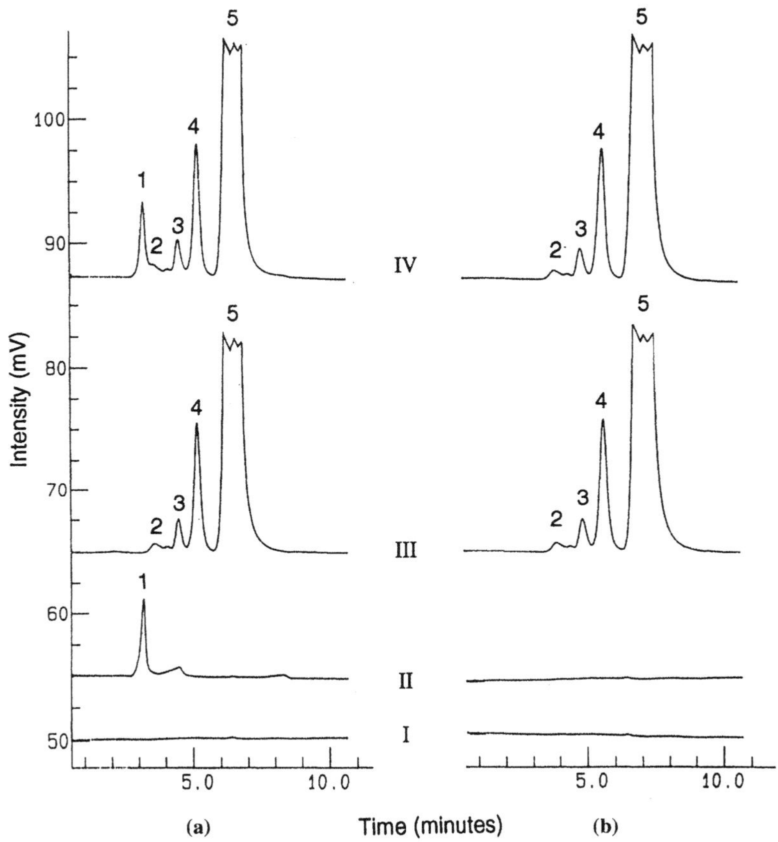

Figure 4.5 In-line dimineralisation. Column, Shodex KS-801 (Na^+) at 70°C. (a) Without in-line deashing; (b) with Bio-Rad in-line deashing. Solvent, water at 1.0 ml min^{-1}. Detection, differential refractometer. (I) Bidistilled water, (II) sodium chloride (20 mg l^{-1}), (III) high dextrose syrup, (IV) high dextrose syrup spiked with 20 ppm of sodium chloride. 1, Chlorides, 2, polysaccharides; 3, maltotriose; 4, maltose; 5, dextrose.

other with an anion exchange resin, are introduced into a cartridge holder which is placed in the HPLC system between the pump outlet and the column inlet. When the sample is injected, prior to reaching the column, it passes through the deashing system, and salts are removed. The effect is illustrated in Figure 4.5. There are other cartridge systems available but work in our laboratory showed that the Bio-Rad cartridges had the highest capacity for salts removal. Another important point to verify is that no adsorption of carbohydrate species occurs.

(c) Filtration. The diluted glucose syrup should be filtered through a 0.22 or 0.45 μm filter before injection. This prevents premature blocking

of the column frits, fouling of the column and increase of the back pressure.

4.3.4.5 Standardisation. Any analysis will only be as good as standardisation permits and primary standards must be made to validate the HPLC system. Several ways of standardisation, or calibration, are often used in chromatography. The simplest for glucose syrups analyses is the area percent method, described here.

To prepare primary standards, pure substances must be employed, and in the case of sugars, several points have to be highlighted:

- The sugar must be dry, or its moisture content must be determined by one of the methods described in section 4.2.
- The sugar must be pure, or its purity must be known. This is one of the weak points of the system, as the best technique to determine the purity is probably HPLC. Indeed, our efforts have shown that verifying HPLC by a second method, e.g. GLC or enzymic, is inadequate due to a higher tolerance. A system with good resolution has to be chosen to make the purity determination, and this is inevitably HPLC itself. The use of a polymer-based cation exchange column in the calcium form, where all gluco-oligomers have very close response factors, should be used to determine the purity.

For starch hydrolysates, the following reference materials should be available: glucose, maltose, maltotriose and polysaccharides. The latter can contain any oligosaccharide from maltotetraose through very high molecular weight, water-soluble polysaccharides. It should be free from glucose, maltose and maltotriose. It can be obtained by using a low DE maltodextrin, or better, by isolation of the polysaccharides fraction of a corn syrup. For syrups containing other hexoses or pentoses, the sugar purity and the moisture must be established. Typical values for commercial saccharides and for the isolated polysaccharides fraction are given in Table 4.1.

Knowing the purity, and the moisture content of each component, a primary standard, having a composition as close as possible to the sample

Table 4.1 Composition of commercial carbohydrates and of isolated polysaccharides fraction

	Dextrose	Maltose	Maltotiose	Polysaccharides
Dry substance (%)	99.89	94.87	95.52	98.48
Dextrose (%)	99.842	0.191	–	0.165
Maltose (%)	0.158	98.920	–	0.278
Maltotriose (%)	–	0.489	96.306	0.831
Polysaccharides (%)	–	0.400	3.694	98.726

to be analysed, is prepared. An example of the preparation of a primary standard, using the data from Table 4.1, is given in Table 4.2. This standard has a composition close to a high dextrose corn syrup.

A solution of the sugars is made in a 100 ml volumetric flask. Subsequently this can be subdivided, stored in sealed vials, and kept frozen for future use. From the weight of each component, and knowing its analysis, the actual analysis of the composite primary standard is calculated. An example is given in Table 4.3.

An aliquot of the standard is injected several times on the HPLC, peak areas are recorded and area percentages are calculated. When integration parameters are correctly set, then area percent figures correspond closely to the actual composition of the standard. If this is not achieved, area normalisation, using different response factors for each component, can be used. After establishing the primary standard, an actual glucose syrup can be installed as a secondary standard. Using the primary standard to

Table 4.2 Preparation of a primary standard

Component	Dry substance (%)	Purity (HPLC) (%)	Weight as is (%)	Weight dry base (g)	Weight per sugar (g)	Final weight (g)
Dextrose	99.89	99.842 DP 1	9.5657	9.5552	9.54010	9.5408
		0.158 DP 2			0.01510	
Maltose	94.87	0.191 DP 1	0.2030	0.1926	0.00037	0.2062
		98.920 DP 2			0.19051	
		0.489 DP 3			0.00094	
		0.400 DP *n*			0.00077	
Maltotriose	95.52	96.306 DP 3	0.1013	0.0968	0.09322	0.0959
		3.694 DP *n*			0.00036	
Polysaccharides	98.48	0.165 DP 1	0.2133	0.2101	0.00035	0.2086
		0.278 DP 2			0.00058	
		0.831 DP 3			0.00175	
		98.726 DP *n*			0.20742	
				Total weight		10.0515

Table 4.3 Primary standard composition

Component	Weight (g)	Theoretical values (%)
Dextrose	9.5408	94.92
Maltose	0.2062	2.05
Maltotriose	0.0959	0.95
Polysaccharides	0.2086	2.08
Total	10.0515	100.00

standardise the system, the syrup is injected at least three times or more to obtain a repeatable result. The analysis established, this type of standard can be used for most internal analyses. The advantage of this approach is convenience with regard to sample availability and suitability.

Apart from the area percent method described above, other calibration methods exist and can be used when the area percent method is not applicable. The more current methods are area normalisation, external calibration and internal calibration. All these methods are often fully described in the reference manual accompanying any data acquisition system.

The calibrations are often based on peak height or peak area. When the HPLC system is calibrated, quantification is possible. Again, these procedures, depending on how the instrument was calibrated, are described in the reference manual of the data processing system and will not be detailed here.

4.3.4.6 Detection. The principal detection system for the HPLC of sugars is the differential refractometer (DRI), a universal mass detector. While several types exist, the majority use the deflection principle which measures the refraction of a parallel beam crossing an interface that separates two media of different refractive index. Modern instruments are now refined to give high sensitivity. Another interesting approach is the interferometric refractive index detector which measures the speed of light in a sample cell and a reference cell by the interference of the two light beams after passage through the cell. This detector is extremely sensitive and is best used on a system that remains uninterrupted.

One of the disadvantages of the DRI is that changes in flow affect its response, inhibiting the use of gradient flow. This problem can be overcome by the evaporative light scattering detector (ELSD) which is also a mass detector developed by Macrae and Dick (1981). The effluent enters the detector and is nebulised by a high velocity stream of nitrogen. The volatiles present are evaporated as the mist is carried through a heated tube forming an aerosol of fine particles. These particles enter the light path of a laser beam and cause scattering of the light which is detected by a photodiode or a photomultiplier. The degree of scattering is measured and can be correlated with the mass of analyte that entered the detector. With the ELSD, Lafosse *et al.* (1992) showed that the potential of silica columns can be further exploited for oligomer analysis, using methanol as solvent with gradient elution.

The ELSD will find application mostly in research, as detector working conditions must be well defined for each type of analysis. Because of its ease of use and suitability the DRI will remain the most used detector for carbohydrates. The electrochemical detector is treated separately in the section on anion exchange chromatography.

4.3.4.7 Cation exchange chromatography.

(a) General applications. Due to their ease of use, cation exchange columns are the 'work horses' for the analysis of glucose syrups. However, there are some separations where modified silica columns are better, and this is so for the resolution of DP 2 isomers such as maltose, isomaltose and maltulose as already shown in Figure 4.4. Their separation is also possible with anion exchange chromatography (as described later).

The most common requirement is the sugar balance in syrups of all kinds, and this is achieved by using cation exchange columns either in the potassium, sodium or calcium form. Best separation of glucose, maltose and maltotriose is obtained with either potassium or sodium form, and a typical chromatogram depicting the chromatographic profile of a high dextrose syrup is shown in Figure 4.6. In this instance, sodium was the counterion, as it was also for the analysis of a high maltose-containing syrup in Figure 4.7.

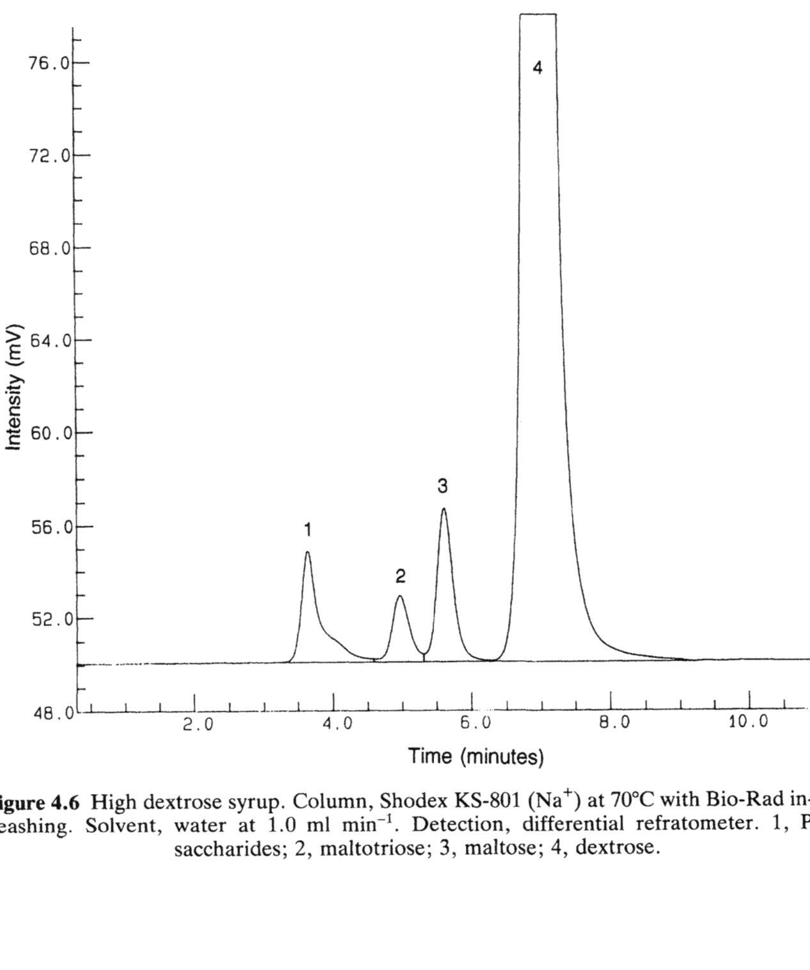

Figure 4.6 High dextrose syrup. Column, Shodex KS-801 (Na^+) at 70°C with Bio-Rad in-line deashing. Solvent, water at 1.0 ml min^{-1}. Detection, differential refratometer. 1, Polysaccharides; 2, maltotriose; 3, maltose; 4, dextrose.

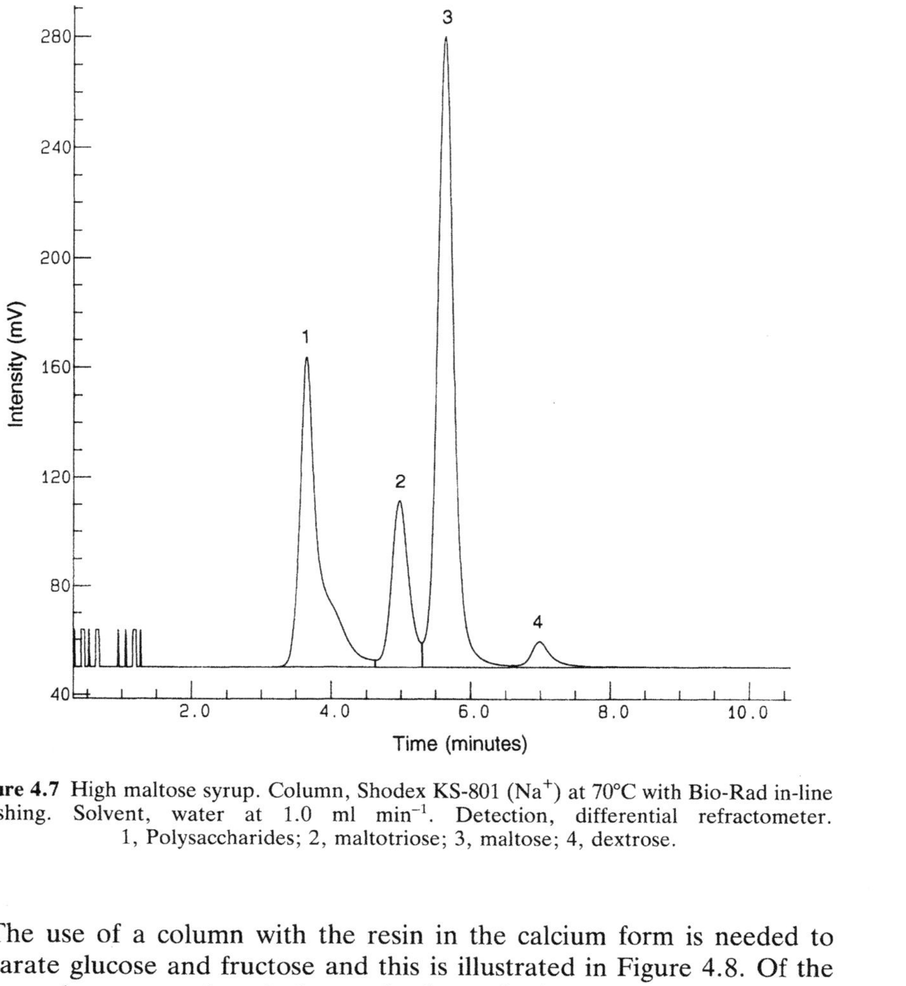

Figure 4.7 High maltose syrup. Column, Shodex KS-801 (Na^+) at 70°C with Bio-Rad in-line deashing. Solvent, water at 1.0 ml min^{-1}. Detection, differential refractometer. 1, Polysaccharides; 2, maltotriose; 3, maltose; 4, dextrose.

The use of a column with the resin in the calcium form is needed to separate glucose and fructose and this is illustrated in Figure 4.8. Of the three columns mentioned above, the latter is the best general column, because of its ability to handle syrups of all types. The use of a column in the hydrogen form can also be used for glucose syrup analysis and an example is depicted in Figure 4.9.

The need for in-depth analysis of glucose syrups reflects the sophistication of both their manufacture and the customer's technology. Analysis of the oligosaccharide fraction is often a requirement and this need is satisfied by use of a column with the resin in the silver form. Of course, the exclusion limit of the column is determined by the degree of crosslinking of the resin, and calcium can also be used. In Figure 4.10, a comparison of the chromatographic profiles obtained on resins with 4% crosslinking is shown. We use a column in the silver form, to obtain the best profile in the shortest time. It is obligatory to use in-line demineralisation with this

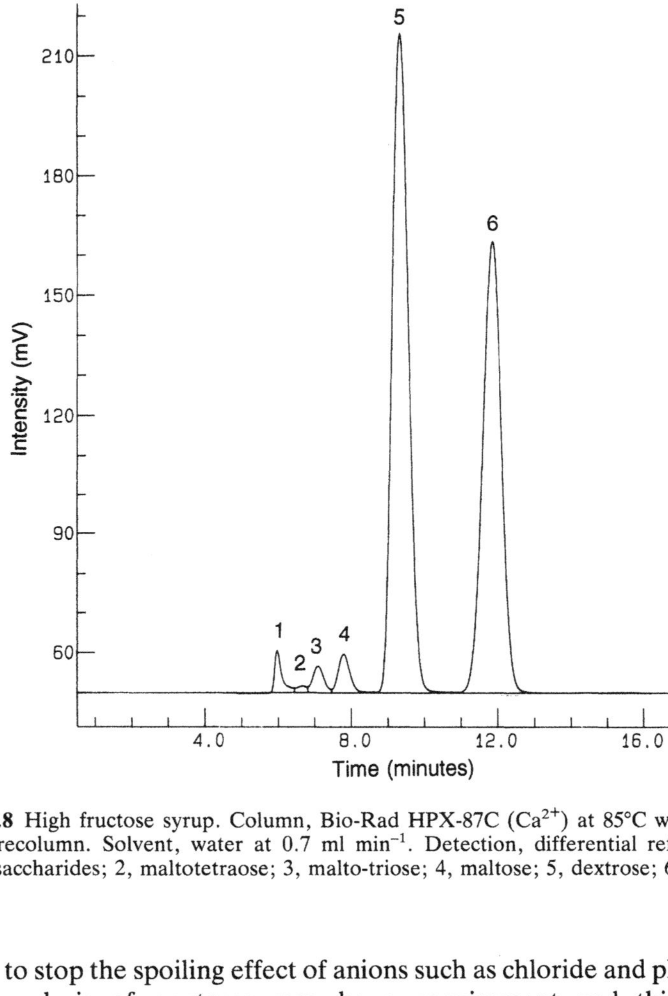

Figure 4.8 High fructose syrup. Column, Bio-Rad HPX-87C (Ca^{2+}) at 85°C with Bio-Rad Carbo precolumn. Solvent, water at 0.7 ml min^{-1}. Detection, differential refractometer. 1, Polysaccharides; 2, maltotetraose; 3, malto-triose; 4, maltose; 5, dextrose; 6, fructose.

column to stop the spoiling effect of anions such as chloride and phosphate.

The analysis of pentoses may be a requirement and this can be performed on a cation exchange column in the lead form. This is illustrated in Figure 4.11. In-line demineralisation is also essential for the same reasons as for the silver form.

Sugar alcohols are well defined when a column in the calcium form is employed but even better resolution is obtained when the lead column is used, as shown in Figures 4.12 and 4.13. The lead column also separates pentitols as illustrated in Figure 4.14. A drawback of the latter column is that often analysis time is quite long and that it is not an easy form to regenerate.

The presence of minor components in glucose syrups, such as organic acids and alcohols can be determined using the column in the hydrogen form. For organic acids in particular, high sensitivity is possible by the use

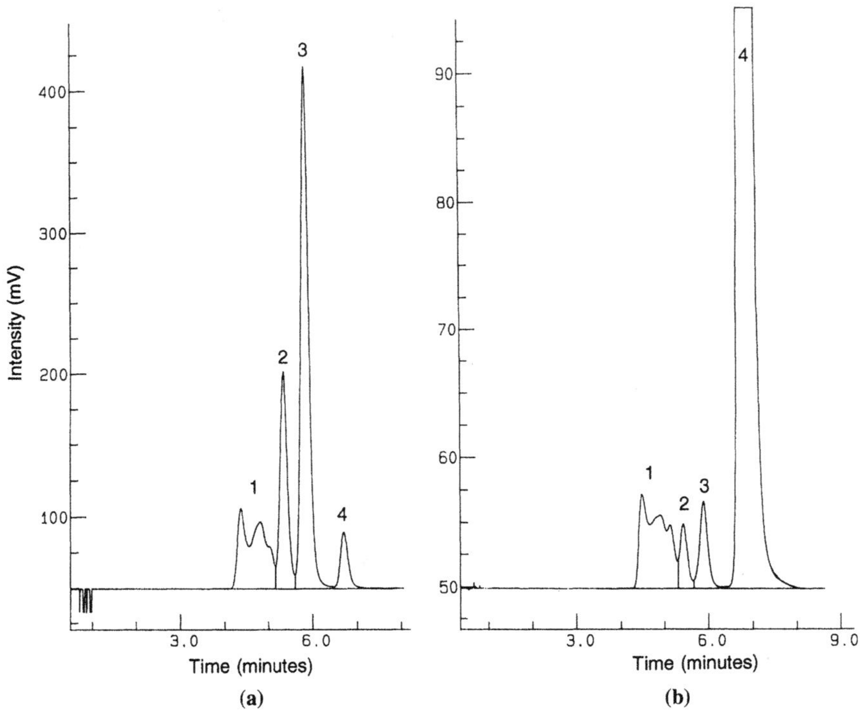

Figure 4.9 Glucose syrups. (a) High maltose syrup, (b) high dextrose syrup. Column, Shodex KC-811 (H^+) at 70°C. Solvent, H_2SO_4 0.01% v/v in water at 1.0 ml min^{-1}. Detection, differential refractometer. 1, Polysaccharides; 2, maltotriose; 3, maltose; 4, dextrose.

of ultraviolet detection at 210 nm, and this is demonstrated in Figure 4.15. Also the interference of the saccharides is reduced to a strict minimum.

Over the years many sugars have been analysed on a wide range of columns and our own particular findings are summarised in Table 4.4. The relative retention times of a wide set of carbohydrates are reported for different columns and have been calculated against the retention time of D-glucose. Table 4.4 gives immediately a good idea of the selectivity of a column for selected components.

(b) Characterisation of higher polymers. Each method for hydrolysing starch gives a specific profile of oligo- and polysaccharides in the final syrup. Characterisation of the higher molecular weight fraction is of interest, whether it is less than 1% in a high dextrose syrup or the major part of a low DE syrup. Size exclusion columns are now available which are suited to elution of polysaccharides. The important criteria are non-adsorption of the solute on the column, a good resolution of the species present and a trusted calibration. Good experience has been found using

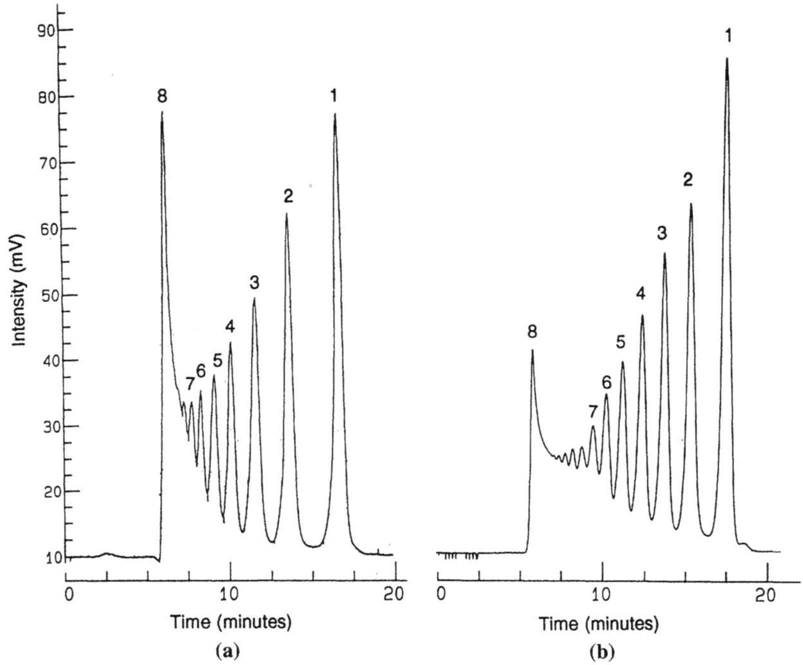

Figure 4.10 Oligosaccharides resolution in acid glucose syrup. (a) Column, 30 cm packed with Benson BC-X4 resin, 10–15 μm in the Ca^{2+} form, at 80°C. Solvent, water at 0.4 ml min^{-1}. Detection, differential refractometer (b) Bio-Rad HPX-42A (Ag^{+}) at 85°C with Bio-Rad in-line deashing. Solvent, water at 0.6 ml min^{-1}. Detection, differential refractometer. 1, Dextrose; 2, maltose; 3, maltotriose; 4, maltotetraose; 5, maltopentaose; 6, maltohexaose; 7, maltoheptaose; 8, polysaccharides.

the Shodex S-800 series of columns, which are cation exchange columns in the sodium form. The series of six columns has exclusion limits from 10^3 to 5×10^7 Da. For glucose syrups, the S-803 column with an exclusion limit of 5×10^4 Da is adequate. Calibration of the system is best effected with a set of pullulans from Shodex P-800, with well defined molecular weights and a narrow distribution. Their preparation is fully described by Kawahara and Ohta (1983). The inclusion of an S-801 column in series with an S-803, provides a better window. Experience has shown that different packings will give different resolution and that results should always be qualified by quoting the chromatographic system used. Good examples of size exclusion profiling of two different syrups are given in Figures 4.16 and 4.17.

Matrix assisted laser desorption ionisation (MALDI) is a technique with a growing range of applications for analysis of biomolecules including carbohydrates. Certain matrices will absorb energy from a laser beam and

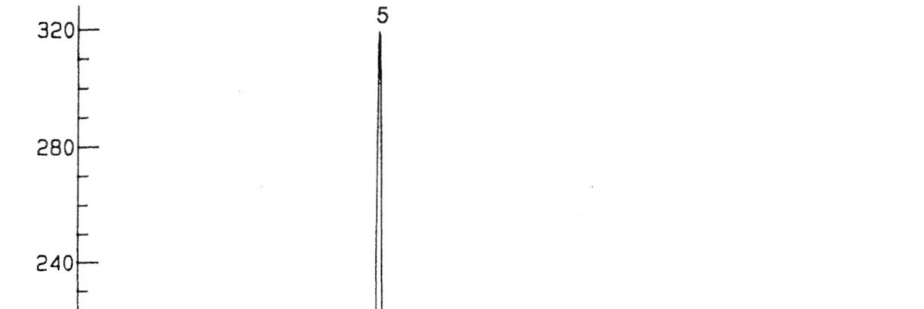

Figure 4.11 Pentoses and hexoses. Column, Bio-Rad HPX-87P (Pb^{2+}) at 85°C with Bio-Rad in-line deashing. Solvent, water at 0.4 ml min^{-1}. Detection, differential refractometer. 1, Polysaccharides; 2, maltose; 3, maltulose; 4, dextrose; 5, xylose; 6, galactose; 7, arabinose; 8, mannose; 9, idose; 10, ribose.

will then assist in desorption and ionisation of biomolecules in the vicinity. The ionised molecules then drift down a tube, where their time of flight is measured and can be related to molecular mass, as reported by Metzger *et al.* (1994). The method is fast and needs minimal sample, but as yet is not truly quantitative. However, it is a fast moving technology.

(c) Pentosans. The use of wheat starch for glucose syrup manufacture can introduce minor constituents not usually found in other starches. One of these in particular is pentosan which might cause haze in the final syrup. Measurement of free pentoses can be done with cation exchange columns in the lead form as described elsewhere in this chapter. There are several ways to measure pentosans, the most frequently used being that based on acid hydrolysis of the pentosans to pentoses followed by condensation of

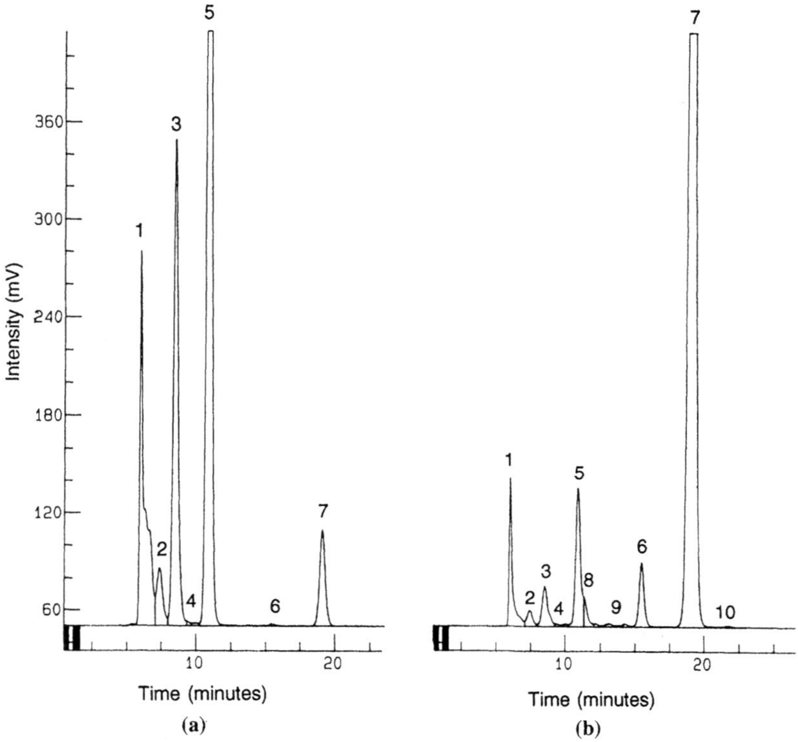

Figure 4.12 Sugar alcohol syrups. (a) Maltitol syrup, (b) sorbitol syrup. Column, Bio-Rad HPX-87C (Ca^{2+}) at 85°C with Bio-Rad Carbo pre-column. Solvent, water with 0.7 ml min^{-1}. Detection, differential refractometer. 1, Polyols; 2, maltotetraitol; 3, maltotriitol; 4, glucopyranosyl-(1-6)-mannitol; 5, maltitol; 6, mannitol; 7, sorbitol; 8, glucopyranosyl-(1-6)-sorbitol; 9, cracking products; 10, iditol.

the pentoses to furfural. The furfural is removed continually by distillation over a 2 h period after which it can be determined by GLC, colorimetrically, or perhaps best by HPLC. It is a time consuming method that is difficult to automate. A more rapid empirical method has been proposed by Douglas (1981) based on hydrolysis in a sealed tube, with colorimetric detection of the furfural formed.

An alternative unreported empirical method has been developed which is also based on hydrolysis and subsequent condensation of the resulting pentoses in a sealed tube. The furfural is quantified on HPLC using UV detection at 283 nm. Calibration is achieved by subjecting xylose and arabinose to the same procedure. A response factor is calculated assuming that the average ratio of xylose:arabinose is 62:38 and that arabinose dehydrates to furfural at a slower rate than xylose. Very good correlation with the distillation method has been found. This method is suitable for all

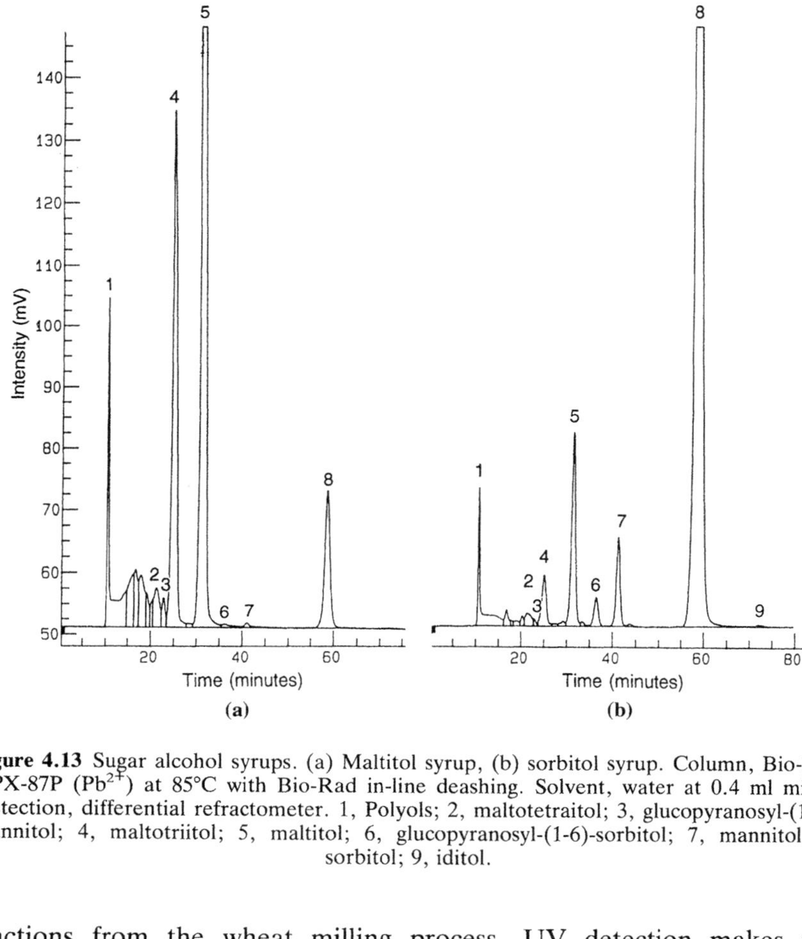

Figure 4.13 Sugar alcohol syrups. (a) Maltitol syrup, (b) sorbitol syrup. Column, Bio-Rad HPX-87P (Pb^{2+}) at 85°C with Bio-Rad in-line deashing. Solvent, water at 0.4 ml min^{-1}. Detection, differential refractometer. 1, Polyols; 2, maltotetraitol; 3, glucopyranosyl-(1-6)-mannitol; 4, maltotriitol; 5, maltitol; 6, glucopyranosyl-(1-6)-sorbitol; 7, mannitol; 8, sorbitol; 9, iditol.

fractions from the wheat milling process. UV detection makes this determination very sensitive and the basic HPLC methodology is also used for other furan compounds (e.g. 5-hydroxymethylfurfural) in glucose syrups.

(d) Total carbohydrate. The total carbohydrate content of a glucose syrup, assuming that it contains only glucose and its polymers, can be measured by various means. In the first step the syrup is hydrolysed fully to glucose, either with enzymes or with acid. The glucose can be quantified either by HPLC or by an enzymatic analysis.

Hydrolysis of the polymers with enzymes is best achieved using a mix of enzymes. It has been claimed that amyloglucosidase alone effects total hydrolysis but work done in our laboratory proves that a limit dextrin remains. The inclusion of a small amount of pullulanase is effective in breaking down this limit dextrin. About 250 mg of sample is dissolved in

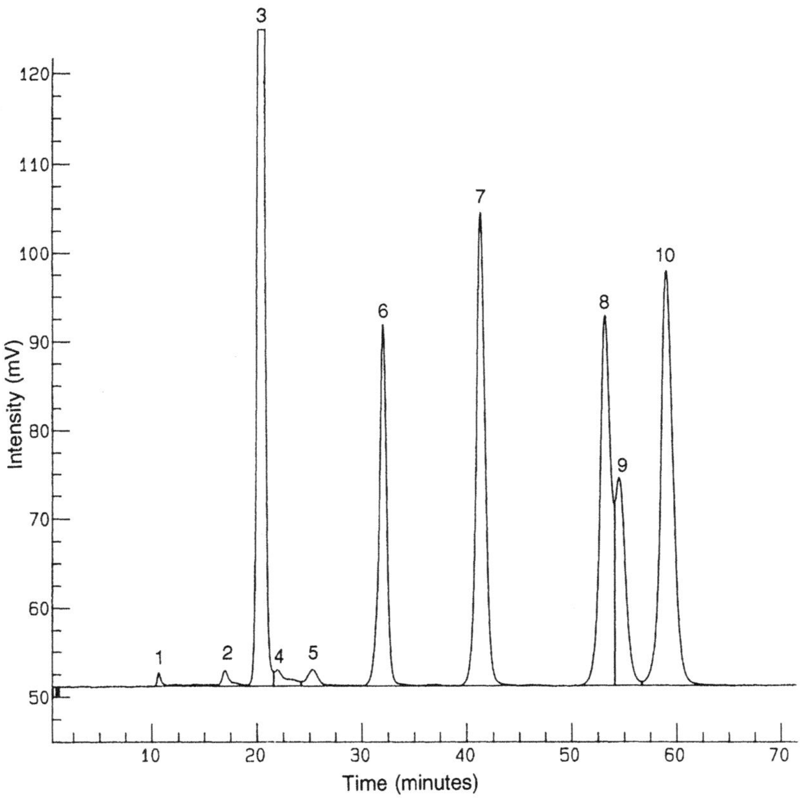

Figure 4.14 Pentitols, pentoses, hexitols and hexoses. Column, Bio-Rad HPX-87P (Pb^{2+}) at 85°C with Bio-Rad in-line deashing. Solvent, water at 0.4 ml min^{-1}. Detection, differential refractometer. 1, Polyols; 2, maltose; 3, dextrose; 4, xylose; 5, arabinose; 6, ribitol; 7, arabitol; 8, xylitol; 9, galactitol; 10, sorbitol.

50 ml of water, 2 g of sodium acetate is added and the pH is adjusted to 4.8 to 5.2. To this is further added pure amyloglucosidase and a minor amount of pure pullulanase. The test solution is then incubated at 60°C for 45 min, after which 10 ml of a 15% solution of sulphuric acid is added to deactivate the enzymes. After cooling and making up to volume, a filtered aliquot is injected on a cation exchange column in the hydrogen form. External calibration is usually adequate, but an internal standard such as meso-erythritol can be introduced, as indicated in Figure 4.18.

For acid hydrolysis an adaptation of an EC approved method (EC, 1987) for dextrins has proved successful. The carbohydrate substrate is refluxed with dilute acid for 3 h; analysis has shown the hydrolysis is complete. In the original method measurement of the glucose is done by the Lane–Eynon titration, but a better way is by HPLC using the same procedure as

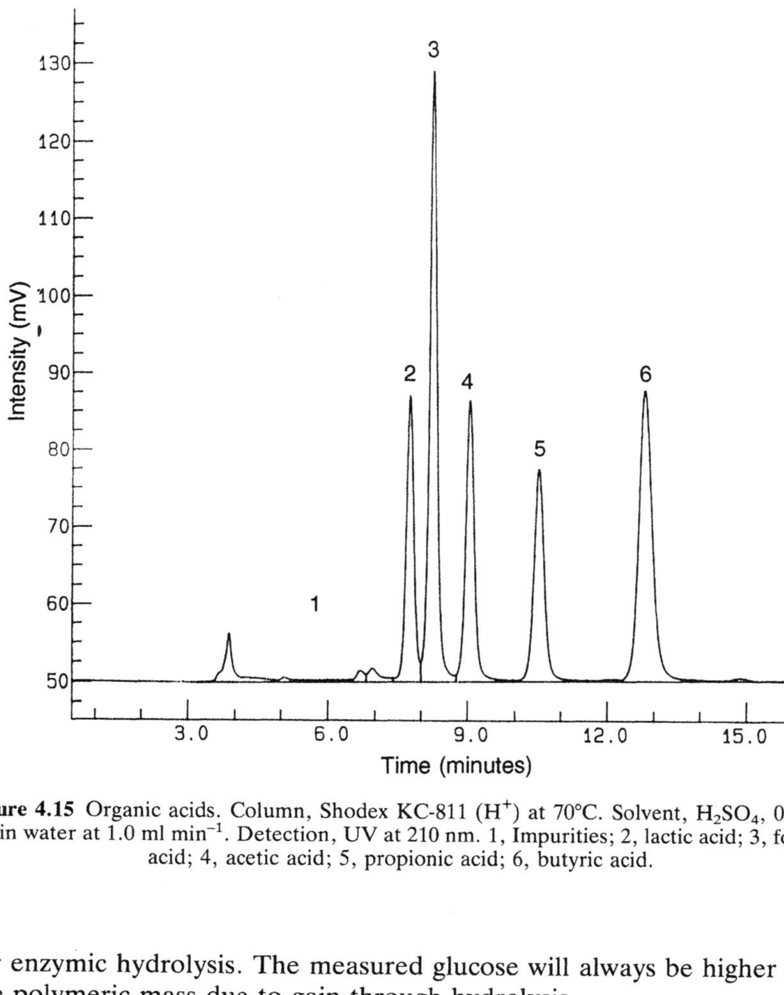

Figure 4.15 Organic acids. Column, Shodex KC-811 (H^+) at 70°C. Solvent, H_2SO_4, 0.01% v/v in water at 1.0 ml min^{-1}. Detection, UV at 210 nm. 1, Impurities; 2, lactic acid; 3, formic acid; 4, acetic acid; 5, propionic acid; 6, butyric acid.

for enzymic hydrolysis. The measured glucose will always be higher than the polymeric mass due to gain through hydrolysis.

(e) Column regeneration. An economic advantage of ion exchange resins over silica is that they can be regenerated and reused. After removal from the column, the resin is treated with 2 M acid to convert it to the acid form. Generally hydrochloric acid is used, but for resins with a heavy metal counterion, nitric acid must be used. Subsequently the acid is removed by thorough washing and the resin is rinsed with a salt solution containing the desired counterion. Nitrates have to be used for the heavy metal ions, but chlorides can be used for others. When all the hydrogen ions are replaced, the resin is washed clear of electrolyte solution and is ready to be repacked. Some loss is inevitable so a quantity of spare resin is essential. Either a specially designed pump can be used for repacking or a conventional

Table 4.4 Relative retention times of saccharides versus the retention time of glucose on five different cation exchange columns

	Shodex KS-801 Na^+	Shodex KC-811 H^+	Bio-Rad HPX-87C Ca^{2+}	Bio-Rad HPX-87P Pb^{2+}	Bio-Rad HPX-42A Ag^+
Allose	1.092	1.053	1.515	1.710	1.077
Altrose	1.071	1.039	1.154	1.190	1.052
Arabinose	1.186	1.128	1.263	1.250	1.111
Cellobiose	0.781	0.852	0.809	0.826	0.864
Erythrose	1.322	1.221	1.969	1.104	1.103
Erythrulose	1.423	1.230	1.502	1.544	1.104
Fructose	1.101	1.062	1.245	1.350	1.069
Fucose	1.154	1.176	1.251	1.250	1.159
Galactose	1.073	1.049	1.116	1.150	1.072
Gentiobiose	0.771	0.837	0.789	0.796	0.854
Glucose	1.000	1.000	1.000	1.000	1.000
Gulose	1.079	1.050	1.449	1.500	1.075
Idose	1.145	1.065	1.282	1.480	1.101
Isomaltose	0.805	0.850	0.817	0.863	0.888
Isomaltotriose	0.708	0.787	0.733	0.775	0.797
Lactose	0.808	0.878	0.856	0.910	0.927
Lyxose	1.120	1.098	1.312	1.310	1.071
Maltitol	0.836	0.904	1.168	1.726	0.951
Maltose	0.801	0.856	0.833	0.878	0.883
Maltotetraose	0.690	0.737	0.664	0.780	0.740
Maltotriose	0.718	0.781	0.756	0.818	0.799
Maltulose	0.809	0.871	0.877	0.924	0.896
Mannitol	1.009	1.079	1.642	2.276	1.053
Mannose	1.080	1.048	1.197	1.310	1.072
Melibiose	0.809	0.860	0.848	0.953	0.914
Palatinose	0.793	0.855	0.848	0.891	0.881
Panose	0.691	0.794	0.737	0.783	0.788
Psicose	1.161	1.088	1.846	2.590	1.084
Raffinose	0.673	–	0.749	0.775	–
Rhamnose	1.028	1.102	1.149	1.170	1.050
Ribose	1.340	1.164	2.012	2.620	1.171
Ribulose	1.280	1.151	1.952	1.608	1.102
Saccharose	0.792	–	0.824	0.828	–
Sorbitol	1.053	1.098	2.005	2.900	1.076
Sorbose	1.031	1.022	1.126	1.150	1.013
Talose	1.312	1.119	1.835	2.950	1.192
Trehalose	0.777	0.864	0.821	0.841	0.864
Turanose	0.796	0.871	0.849	0.910	0.885
Xylose	1.090	1.057	1.102	1.080	1.032
Xylulose	1.137	1.102	1.281	1.305	1.056

HPLC pump suffices. For both, an upper pressure limit of 1500 psi is advised. New frits should be installed, or thorough cleaning of the used ones with nitric acid.

Packed columns have been obtained by the technique described with at least equal efficiency to a purchased column. Regeneration can be repeated many times, our own experience showing that ten times or more

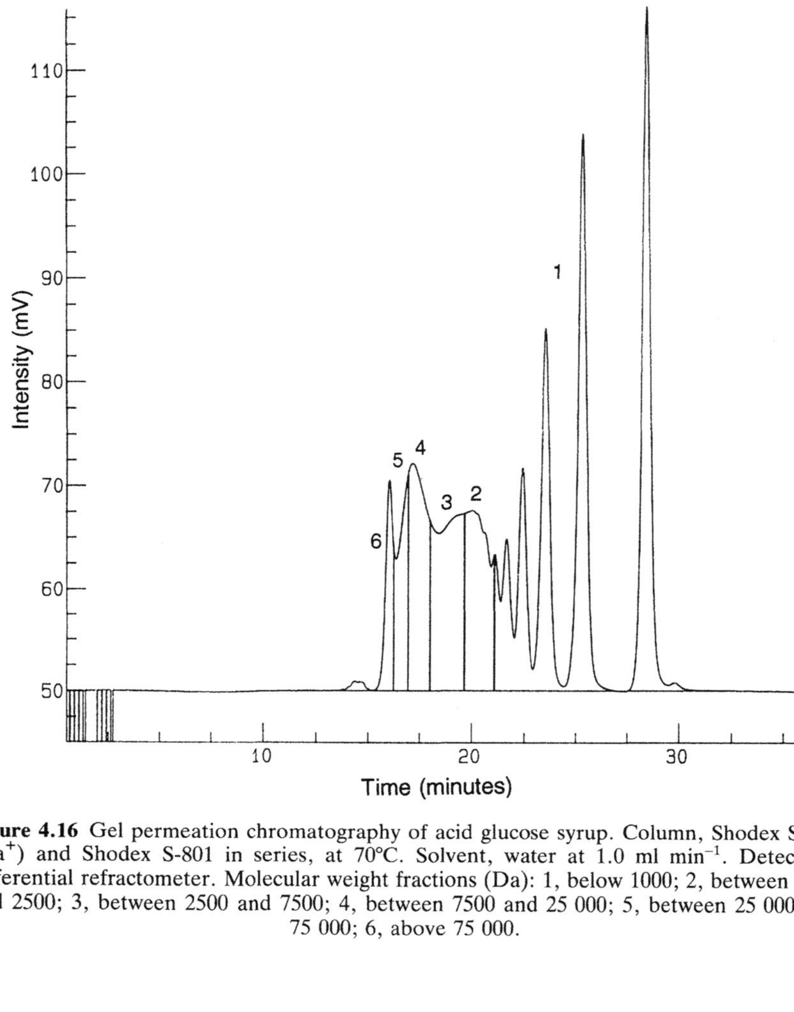

Figure 4.16 Gel permeation chromatography of acid glucose syrup. Column, Shodex S-803 (Na^+) and Shodex S-801 in series, at 70°C. Solvent, water at 1.0 ml min^{-1}. Detection, differential refractometer. Molecular weight fractions (Da): 1, below 1000; 2, between 1000 and 2500; 3, between 2500 and 7500; 4, between 7500 and 25 000; 5, between 25 000 and 75 000; 6, above 75 000.

is no exception. Our experience with *in situ* regenerations is limited because successful regeneration was never obtained, probably due to the swelling properties of the resin at different pH values and in contact with various electrolytes.

4.3.4.8 Anion exchange chromatography

(a) Introduction. The term 'ion chromatography' for carbohydrate analysis needs some explanation. Carbohydrate analysis using cation exchangers as separation media is the major technique used within the industry. However, a better description of 'ion chromatography' for carbohydrates could be chromatographic carbohydrate analysis on anion

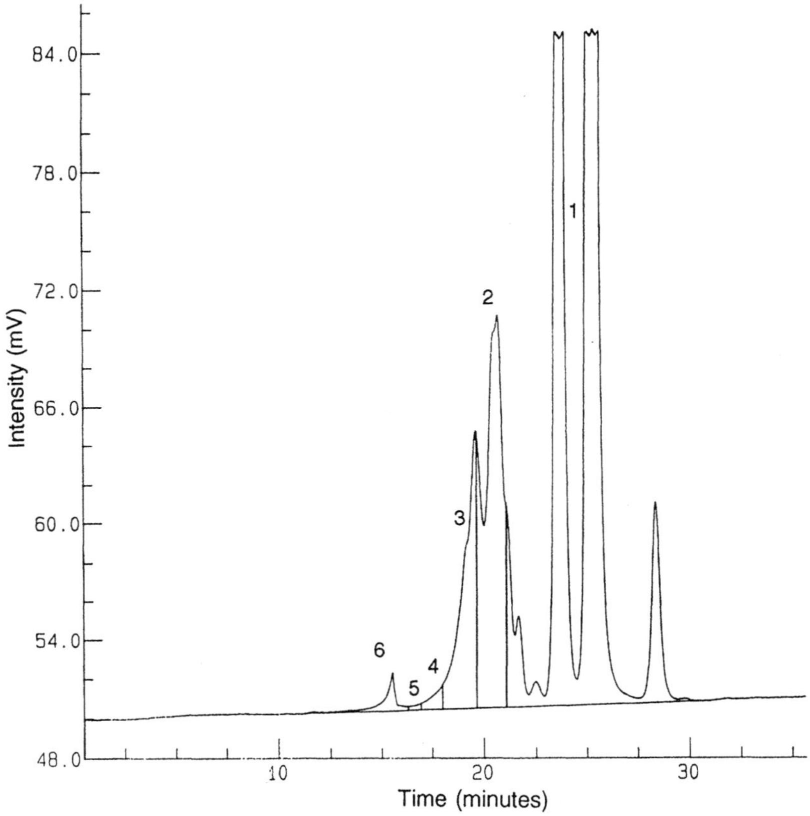

Figure 4.17 Gel permeation chromatography of high maltose syrup. Column, Shodex S-803 (Na^+) and Shodex S-801 in series, at 70°C. Solvent, water at 1.0 ml min^{-1}. Detection, differential refractometer. Molecular weight fractions (Da): 1, below 1000; 2, between 1000 and 2500; 3, between 2500 and 7500; 4, between 7500 and 25 000; 5, between 25 000 and 75 000; 6, above 75 000.

exchanger columns using electrochemical detection. Indeed, the major benefits of using anion exchanger columns with electrochemical detection reside in the remarkable and flexible separation powers of the system and in the very high sensitivity of the detection. The specificity of the detection mode permits gradient elution, providing a whole gamut of separations.

Ion chromatography, as its name says, was initially developed for the analysis of ions, e.g. mineral anions, cations, organic acids, amines, amino acids. One company, namely the Dionex Corporation, developed a sophisticated column technology and electrochemical detection technique for it. Besides the already well known conductivity detector, the (pulsed) amperometric detector was made commercially available. Considering

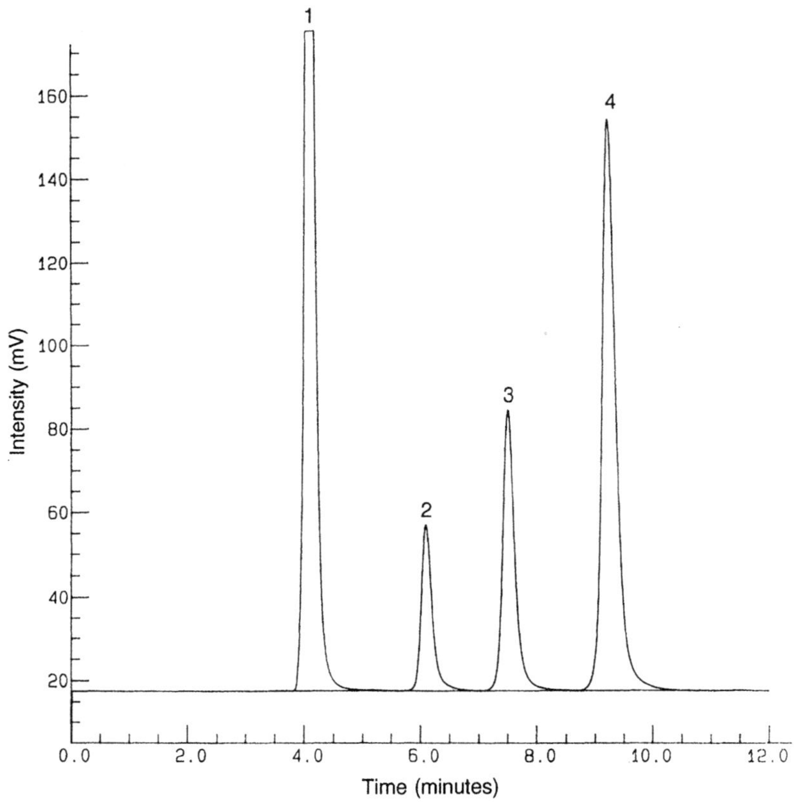

Figure 4.18 Total carbohydrate analysis after enzymic treatment. Column, Shodex KC-811 (H^+) at 70°C. Solvent, H_2SO_4, 0.01% v/v in water at 1.0 ml min^{-1}. Detection, differential refractometer. 1, Sulphuric acid; 2, dextrose; 3, internal standard (erythritol); 4, acetic acid.

carbohydrates as very weak acids, they will behave like anions under strong alkaline conditions. Amperometric detectors can be adjusted in such a way that they become specific for a given analyte. From cyclovoltammetry, parameters can be set to detect carbohydrates at very low levels. Where in conventional carbohydrate HPLC, the use of gradient elution could not be applied with a differential refractometer, the amperometric detection allows gradients and therefore greatly expands the possibilities for use. In combination, the column technology and the detection system make ion chromatography a very powerful analytical tool for carbohydrates in general. Spruyt (1994) overviews miscellaneous techniques for carbohydrate analysis, including ion chromatography.

(b) Columns. The typical anion exchanger packings used for carbohydrate analysis, developed by Small *et al.* (1975), and used by Dionex are

resins of polystyrene/divinylbenzene, surface sulphonated, with completely aminated porous latex particles fixed onto the resin. The latex particles (about 0.1 μm) are agglomerated with the core resin particles (10–20 μm or smaller) and carry the actual anion exchange function ($-N^{+}R_3$). Attachment to the resin core occurs by electrostatic and van der Waals interactions. These latexed anion exchanger resins exhibit enormous chemical stability and the selectivity of the combined system is optimised by varying the quaternary ammonium base as mentioned by Horvath (1974), Weiss (1985, 1986), and Slingsby and Pohl (1988).

Compared to the more 'classic' cation exchanger systems, the anion exchangers show different advantages. The retention times and elution sequences of the carbohydrates can be controlled by varying the ionic strength of the eluent, and by means of this fine tuning, separations of isomers and like substances can be achieved. This is explained by the fact that carbohydrates have pK_a values between 12 and 14. A strongly alkaline eluent can convert the carbohydrates to their anionic forms. By raising the hydroxide concentration in the eluent using gradient systems, even higher oligosaccharides can be analysed. Koizumi *et al.* (1988) mention separation of oligosaccharides up to DP 80.

A drawback of the use of strongly alkaline eluents is the high risk of carbonation. The carbonate is a stronger eluent than the hydroxyl ion and even traces of carbonate give loss of resolution or strong alteration of the elution order. In particular cases, the controlled carbonation can be used beneficially in order to obtain certain difficult separations. Carbonation often occurs in an uncontrolled way and might not be of interest in obtaining repeatable results. The fear of carbohydrate alteration at high pH, due to the Lobry–de Bruyn–von Ekenstein transformation towards dihydroxyacetone, is negligible when the analysis is performed at room temperature.

Commercially, several different columns are available and each type is almost 'tailor-made' for a specific family of carbohydrates (e.g. sugar alcohols). The Dionex Corporation (1994) provides full information in this field.

(c) Detection. The eluent used for the separation of carbohydrates in ion chromatography is very alkaline. When a cyclic voltammogram of a 0.1 M NaOH solution is recorded using a gold working electrode a typical pattern for the behaviour of this supporting electrolyte is observed. When a small quantity of glucose is added and the cyclic voltammogram is recorded again, a different pattern appears, as illustrated in Figure 4.19.

From this new pattern, the potential for the electrochemical oxidation of glucose can be derived. The current generated by this reaction is proportional to the glucose concentration. In a cell designed for HPLC detection, the specific potential for this reaction is applied. However,

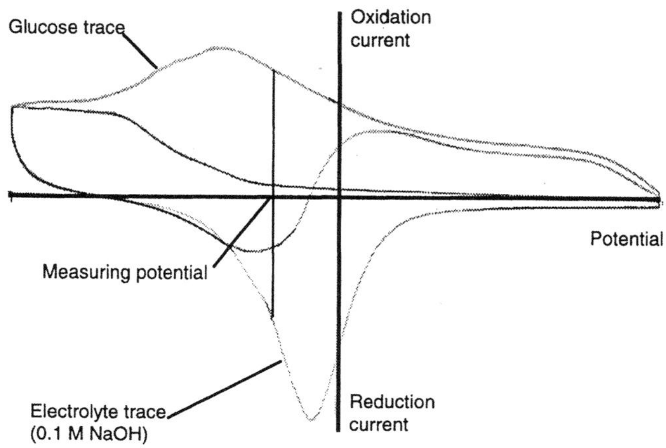

Figure 4.19 Cyclic voltammogram of supporting electrolyte (0.1 M NaOH) and added dextrose, measured against a gold working electrode.

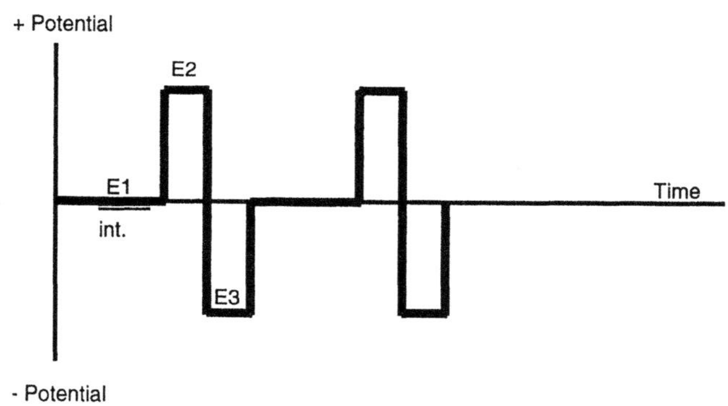

Figure 4.20 General representation of the pulsed potentials in pulsed amperometric detection. *E*1, Measuring potential; *E*2, *E*3, electrode restoring potentials.

polarisation and other phenomena of electrode fouling will rapidly alter the cell response and steps are taken to correct for this. In practice, this is achieved by applying a block-pulsed wave of potentials to the cell and this is repeated in a consecutive cycle in order to regain the original state of the cell electrode.

The measuring potential (*E*1) is used for the actual measurement, the cleaning and restoring potentials (*E*2, *E*3) are used to restore the electrode. Figure 4.20 shows the cycle.

This basic approach is known as pulsed amperometric detection. Further developments make use of an extra variable potential applied at a certain point during the measurement phase. The current variation during this phase is then integrated and this results in better stability of the baseline.

This development is known as integrated amperometry. A full explanation of the mechanism occuring at the electrodes is well described by Larew (1980), Rocklin and Pohl (1983) and Rocklin (1985).

Possible drawbacks of this kind of detector are:

- Limited dynamic range; the high sensitivity of the detection often requires very dilute analyte. If a wide range of concentrations is applied, non-linear response may occur above a certain concentration. This often varies from analyte to analyte.
- Calibration; with differential refractometry the response factors for sugars are very similar, but with amperometric detection this is not necessarily so.
- Maintenance and stability; despite the self-cleaning principle of the pulsed amperometric detectors, regular inspection and mechanical cleaning of the electrodes is required.
- System parameters; for carbohydrate detection, high pH values are needed for proper functioning of the electrochemistry. In some cases post-column addition of an alkali (NaOH) is necessary to maintain full sensitivity and stability throughout the analysis.

(d) Applications

(i) Oligosaccharides. Where in 'classic' cation exchanger HPLC separation of oligosaccharides of glucose polymers can be obtained up to a degree of polymerisation (DP) of 10–15, the anion exchange separation using gradient elution and electrochemical detection can resolve to a much higher DP. A typical application to study starch hydrolysates up to DP 20–25 makes use of an anion exchanger (e.g. Dionex AS6, CarboPac™ PA1, or PA-100). By fine tuning the gradient elution conditions, an even higher DP separation can be obtained.

So as to achieve elution of the higher DP components in a reasonable time, not only should the hydroxyl concentration be increased during the analysis, but also a stronger competing anion should be added to the eluent. Usually acetate is the 'pusher' anion of choice. In Figure 4.21 an example is shown.

More 'state of the art' columns, such as Dionex CarboPac PA1 or PA-100 provide excellent oligosaccharide separations under optimised conditions given by the supplier. It has been observed that low molecular weight oligomers eluted as two peaks, which illustrate the high resolving power of the column in the low concentration period of the gradient. The twin peaks represent the anomers of the carbohydrate (α- and β-form). This phenomenon may, in some cases, make the interpretation of the chromatogram less easy.

(ii) Disaccharide isomers. A typical requirement in glucose manufacture is the quantification of maltose isomers formed by enzyme-

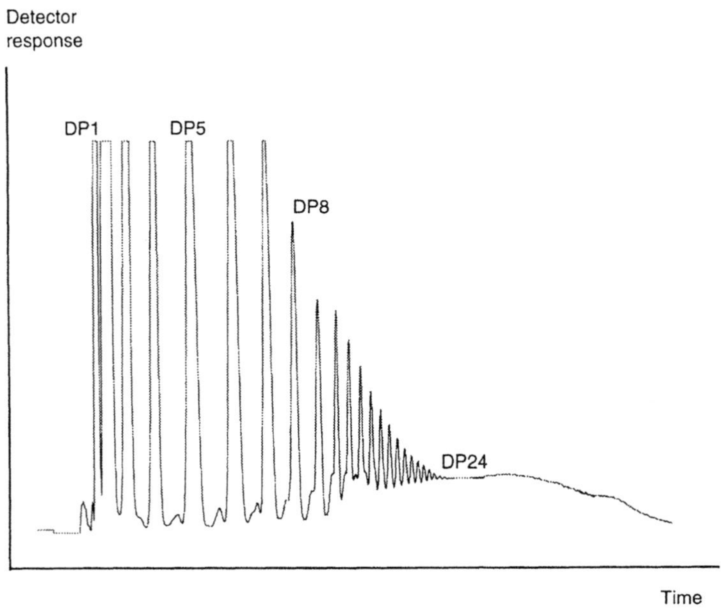

Figure 4.21 Typical chromatogram of oligosaccharides analysed in an acid hydrolysed corn starch. Column, DIONEX IonPac™ AS6.

induced reversion reactions. The analysis can be performed on silica-based columns as previously described but with the inherent disadvantages. A spectacular improvement was obtained using ion chromatography on an anion exchanger using the conditions as for oligosaccharides with NaOH and acetate in the solvent, but at lower concentrations. This is illustrated in Figure 4.22.

Based on these conditions, analysis of other di- and trisaccharides can be done. Modifying the gradient, eleven very similar carbohydrates can be separated to a degree never obtained with any other HPLC columns used presently. Conditions have, of course, to be altered to obtain this kind of separation, which can be seen in Figure 4.23.

As is known, the carbonation of the strongly alkaline eluents occurs very easily with a marked effect on the separation. It has been observed, however, that an involuntary carbonation improved the separation of isomaltose and maltulose. This carbonation cannot be controlled and is thus difficult to reproduce.

(iii) Sugar alcohols. Although the anion exchanger column of the type CarboPac PA1 might be considered for general purpose, special column types have been developed to analyse polyols. The Dionex CarboPac MA1 column is a good example for the separation of reduced mono- and disaccharides. This separation is given in Figure 4.24.

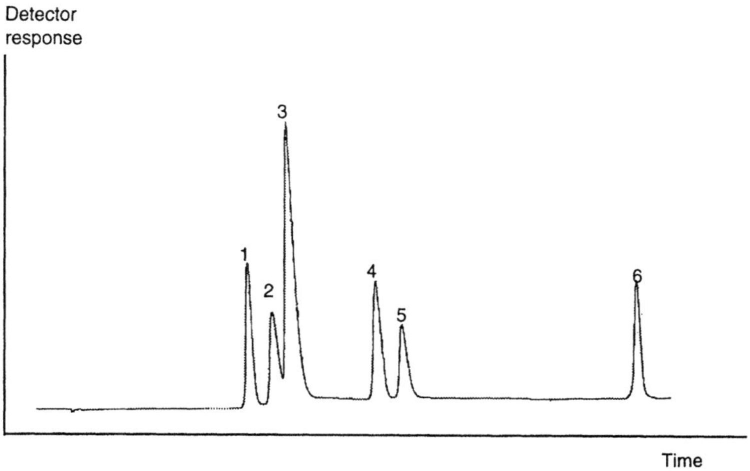

Figure 4.22 Separation of 1, dextrose; 2, fructose; 3, saccharose; 4, isomaltose; 5, maltulose; 6, maltose. Separated on DIONEX IonPac™.

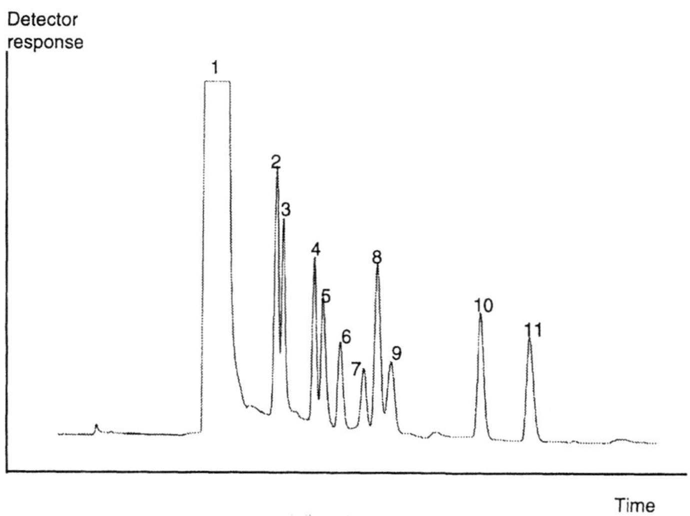

Figure 4.23 Separation of seven DP2 and three DP3 isomers on a DIONEX Ion Pac™ AS6 column. 1, Dextrose; 2, isomaltose; 3, maltulose; 4, gentiobiose; 5, palatinose; 6, isomaltotriose; 7, nigerose; 8, maltose; 9, sophorose; 10, panose; 11, maltotriose.

Being a high efficiency anion exchange column, again the use of isocratic or gradient elution at various hydroxyl concentrations allows flexibility. The system is also suitable for the identification of products of β-elimination. The carbohydrates linked to the β-position of a protein or glycolipid can be analysed after reductive release with $NaBH_4$.

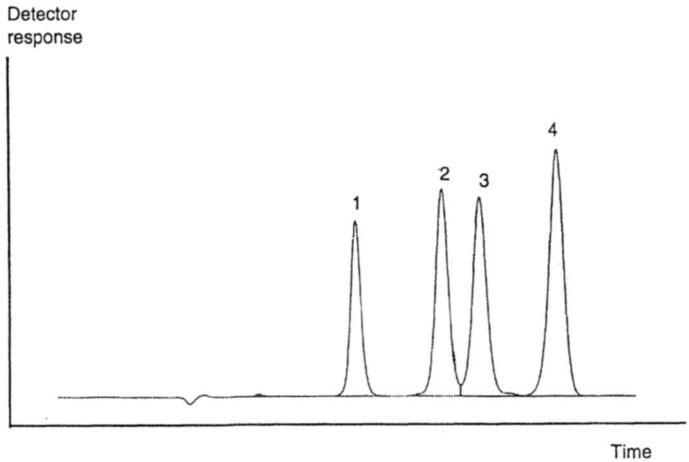

Figure 4.24 Typical application of a dedicated column, using a DIONEX MA1 column. 1, Xylitol; 2, sorbitol; 3, galactitol; 4, dextrose.

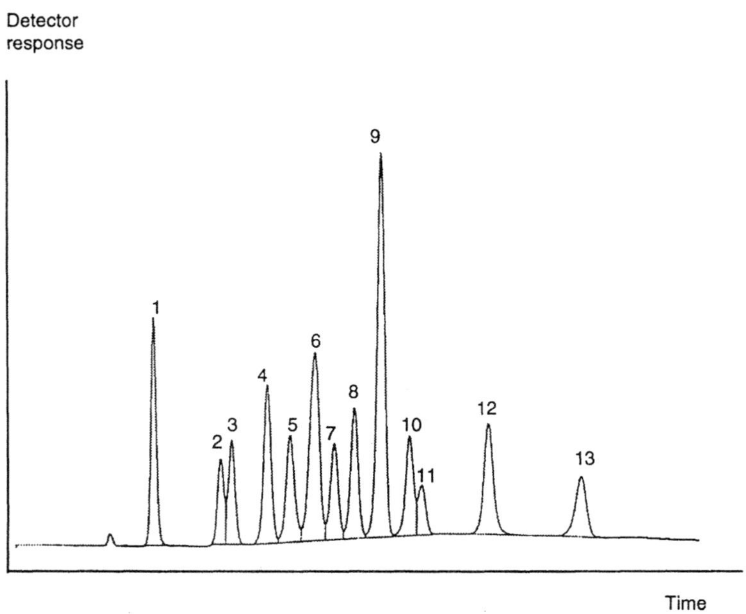

Figure 4.25 Separation on a DIONEX MA1 column under extended conditions. 1, Inositol; 2, xylitol; 3, rhamnose; 4, arabitol; 5, sorbitol; 6, galactitol; 7, mannitol; 8, mannose; 9, dextrose; 10, galactose; 11, fructose; 12, allose; 13, xylulose.

Aminated carbohydrates resulting from this treatment are also separated (e.g. *N*-acetyl glucosaminitol). Certain oligosaccharide alditols released from this or other reactions can be analysed. A problem exists if xylose and glucose are present in the sample as their separation is almost impossible. Restoring the initial conditions and allowing enough time (at least 15 min) to re-equilibrate the column for further analysis are imperative. For best results the concentration of each component should not exceed 20 mg l^{-1}. A typical separation of polyols together with some monosaccharides is given in Figure 4.25.

(e) Conclusions. The use of the latex bonded pellicular anion exchangers combined with electrochemical detection is a valuable complement to classic carbohydrate analysis by HPLC.

Sensitivity and resolution are of a different order and allow the analysis of minor components. Dedicated column types make it possible to study a large number of carbohydrate families. The technique itself, however, requires good knowledge of the separation mechanism and the electrochemical detector needs particular maintenance. It is recommended that in all cases the appropriate precolumns are installed.

References

AFNOR (1986) NF V03–606.
Aitzetmuller, K. (1978) *J. Chromatogr.* **156**, 354.
Bio-Rad (1994) Life Science Research Product, Cat. T. 44.
Boumahraz, M., Davidov, V. and Kiselev, A. (1982) *Chromatographia* **15**, 751.
Bourne, E.J., Tatlow, C.E.M. and Tatlow, J.C. (1950) *J. Chem. Soc.* 1367.
Brobst, K.M., Scobell, H.D., Steele, E.M. (1973) Analysis of carbohydrate mixtures by liquid chromatography, *Proceedings of the 39th Annual Meeting of the American Society Brewing Chemists*, St Paul, Minnesota.
Conrad, E.C. and Palmer, J.K. (1976) *Food Technol.* **30**, (10), 84.
CRA (1959) *Standard Analytical Methods of the Member Companies of the Corn Industries Research Foundation*, 6th edn., 1st revision, Corn Refiners Association, Washington, DC.
CRA (1968) *Standard Analytical Methods of the Member Companies of the Corn Industries Research Foundation*, 6th edn., 1st revision, Corn Refiners Association, Washington, DC.
CRA (1973) *Standard Analytical Methods of the Member Companies of the Corn Industries Research Foundation*, 6th edn., tentative standard, Corn Refiners Association, Washington, DC.
CRA (1975) *Critical Data Tables*, 3rd edn., Corn Refiners Association, Washington, DC.
CRA (1980) *Standard Analytical Methods of the Member Companies of the Corn Industries Research Foundation*, 6th edn., 1st revision, Corn Refiners Association, Washington, DC.
CRA (1983) *Standard Analytical Methods of the Member Companies of the Corn Industries Research Foundation*, 6th edn., 2nd revision, Corn Refiners Association, Washington, DC.
CRA (1985) *Standard Analytical Methods of the Member Companies of the Corn Industries Research Foundation*, 6th edn., 2nd revision, Corn Refiners Association, Washington, DC.
CRA (1988) *Standard Analytical Methods of the Member Companies of the Corn Industries Research Foundation*, 6th edn., 1st issue, Corn Refiners Association, Washington, DC.
CRA (1989a) *Standard Analytical Methods of the Member Companies of the Corn Industries Research Foundation*, 6th edn., 2nd revision, Corn Refiners Association, Washington, DC.
CRA (1989b) *Standard Analytical Methods of the Member Companies of the Corn Industries Research Foundation*, 6th edn., 2nd revision, Corn Refiners Association, Washington, DC.

CRA (1990a) *Standard Analytical Methods of the Member Companies of the Corn Industries Research Foundation*, 6th edn., 1st revision, Corn Refiners Association, Washington, DC.
CRA (1990b) *Standard Analytical Methods of the Member Companies of the Corn Industries Research Foundation*, 6th edn., 1st revision, Corn Refiners Association, Washington, DC.
CRA (1991) *Standard Analytical Methods of the Member Companies of the Corn Industries Research Foundation*, 6th edn., 2nd revision, Corn Refiners Association, Washington, DC.
CRA (1992) *Standard Analytical Methods of the Member Companies of the Corn Industries Research Foundation*, 6th edn., 1st revision, Corn Refiners Association, Washington, DC.
CRA (1993) *Computer Program for the Calculation of Sweetener Physical Properties Tables*, Corn Refiners Association, Washington, DC.
Douglas, S.G. (1981) *Food Chem.* **7**, 139.
E.C. (1987) L 392/25.
Fitt, L.E., Hassler, W. and Just, D.E. (1980) *J. Chromatogr.* **187**, 381.
Fitton, M.G. (1979) *Starch/Stärke* **31**, 381.
Folkes, D.J. and Brookes, A. (1984) *Glucose Syrups: Science and Technology*, Elsevier, p. 214.
Goulding, R.W. (1975) *J. Chromatogr.* **103**, 229.
Heyraud, A. and Rinaudo, M. (1981) *J. Liq. Chromatogr.* **4** Suppl. 2, 175.
Horvath, C. (1974) *Pellicular Ion Exchangers, Bonded Stationary Phases in Chromatography* (ed. E. Grushka), Ann Arbor Science.
ISO (1980) ISO 1742, International Standards Organisation, Geneva.
ISO (1982) ISO 1743, International Standards Organisation, Geneva.
ISO (1983) ISO 5381, International Standards Organisation, Geneva.
ISO (1994) ISO 5809, International Standards Organisation, Geneva.
Kearsley, M.W. (1978) *J. Assoc. Public Analysts* **16**, 1.
Koizumi, K., Kubota, Y., Tanimoto, T. and Okada, Y. (1989) *J. Chromatogr.* **464**, 365.
Koizumi, K., Utamura, T., Kubota, Y. and Hizikuri, S. (1987) *J. Chromatogr.* **409**, 396.
Lafosse, M., Elfakir, C., Morin-Allory, L. and Dreux, M. (1992) *J. High Res. Chromatogr.* **15**, 312.
Larew, L.A. (1988) *Quantitation of Maltooligosaccharides and Determination of Glucoamylase Activity using Pulsed Amperometric Detection*, Dissertation, Iowa State University, Ames, Iowa.
Lochmuller, C. and Hill, W. (1983) *J. Chromatogr.* **264**, 215.
MacLeod, S.K. (1991) *Anal. Chem.* **63**, 557.
Macrae, R. and Dick, J. (1981) *J. Chromatogr.* **210**, 138.
Megazyme (1992) Glucose Assay Kit.
Muller, H. and Siepe, V. (1980) *Chromatographia* **13**, 437.
Robards, K. and Whitelaw, M. (1986) *J. Chromatogr.* **373**, 81.
Rocklin, R.D. and Pohl, C.A. (1983) *J. Liq. Chromatogr.* **6**, 1577.
Rocklin, R.D. (1985) *Adv. Chem. Ser.* **210**, 13.
Sandra, P. and David, F. (1989) *Tenth International Symposium on Capillary Chromatography*, Huethig Verlag, 1045.
Sidisky, L.M., Stormer, P.L., Nolan, L., Keeler, M.J. and Bartram, R.J. (1988) *J. Chromatogr. Sci.* **26**, (7), 320.
Slingsby, R.W. and Pohl, C.A. (1988) *J. Chromatogr.* **488**, 241.
Small, H., Stevens, T.H. and Baumann, W.C. (1975) *Anal. Chem.* **47**.
Spruyt, D. (1994) *Glucose Zetmeel Informatie* **11**, (5), 179.
Sweeley, C.C., Bentley, R., Makita, M. and Wells, W.W. (1963) *J. Am. Chem. Soc.* **85**, 2497.
Verhaar, L., Kuster, B. and Claessens, H. (1984) *J. Chromatogr.* **284**, 1.
Verzele, M., Simoens, G. and Van Damme, F. (1987) *Chromatographia* **23**, (4), 292.
Voragen, A.G.J., Schols, H.A., Searle-van Leeuwen, M.F., Bekdman, G. and Rombouts, F.M. (1986) *J. Chromatogr.* **370**, 113.
Wartman, A.M., Hagberg, C. and Eliason, M.A. (1976) *J. Chem. Eng. Data* **21**, 459.
Weiss, J. (1986) *Handbook of Ion Chromatography*, Dionex Corporation, Sunnyvale, California.
Weiss, J. (1995) *Ion Chromatography*, VCH, Weinheim, Germany.
Wheals, B. and White, P. (1979) *J. Chromatogr.* **176**, 421.
Y.S.I. (1979) Industrial Analyzer Model 27, Yellow Springs Instrument.

5 Physical and chemical properties of glucose syrups

M.W. KEARSLEY and S.Z. DZIEDZIC

5.1 Introduction

The glucose syrup industry has evolved rapidly in recent years to meet the ever increasing demands of the food and related industries. Different types of glucose syrups, glucose/fructose syrups and a multiplicity of glucose syrup, or perhaps more correctly starch, derivatives such as polyols and cyclodextrins, have been introduced to the market. This evolution continues unabated with new products appearing continuously.

Manufacturing techniques, including hydrolysis and refining, undergo continuous improvement to meet the demand for purer products and whilst maize starch was once the main source of starch for these products, manufacturers are now using a much wider range of raw materials. Wheat, potato, tapioca and rice starches are now widely used with potential cost savings over maize depending on the location of the manufacturing site. Whilst the properties of the final glucose syrup are independent of the starch source, the properties of low dextrose equivalant maltodextrins (dextrose equivalent below about 10) are directly related to the type of starch from which they are produced. Thus wheat derived maltodextrins will set to a firm gel whilst potato derived products have a soft spreadable consistency.

The term 'corn syrup' was once widely used to describe these hydrolysis products but the use of different starches has now led to a return to the perhaps more accurate description 'glucose syrup' although they do of course contain glucose polymers in addition to glucose. In the USA corn or maize starch is the predominant starting material for glucose syrup production whilst in Europe maize and wheat are mainly used.

5.2 Dextrose equivalent (DE)

The degree of hydrolysis of the starch, whether by acid or enzyme or a combination of the two, governs the composition of the final product. Complete or 100% hydrolysis of starch gives dextrose, and starch and dextrose are thus the two extremes of the range. The DE scale is therefore related to these two products. Starch has undergone 0% hydrolysis and is thus given a nominal value of 0 DE whilst dextrose, the ultimate hydrolysis product, is given a value of 100 DE. Intermediate values represent the gradual degradation or breakdown of the starch first into large glucose polymers of 20–30 units in size then into much smaller products of 20 units and below. Depending on its DE, a glucose syrup will therefore contain a greater or lesser percentage of high and low molecular weight glucose polymers and it is the percentage of the individual saccharides in a syrup which gives that syrup its unique properties and which differentiates one syrup from another.

Unfortunately it is not simply a matter of determining the DE of a glucose syrup to deduce its properties, as it is possible using different hydrolysis techniques to produce two (or even three) syrups of the same DE with different carbohydrate composition and thus different properties. This is discussed in detail elsewhere in this book. In this chapter however the properties will be related broadly to the DE of the syrup with allowances made where appropriate for obvious discrepancies.

Whilst the scale of DE covers the range from 0 to 100 not all these products are considered as glucose syrups industrially. From 3–20 DE, the products are described as maltodextrins, from 20 to about 75 DE as glucose syrups and above about 75 DE as hydrolysates.

Traditional hydrolysis products from 3–45 DE are generally available as spray dried powders and from 30–95 DE as syrups. Products from 30–45 DE are available in both forms.

It is interesting to note, however, the appearance on the market of spray dried hydrolysates of about 95 DE which traditionally were only available in syrup form. Although industrially glucose syrups are manufactured and sold at a specific DE it is not usually possible for a manufacturer to produce a syrup of the same DE on every occasion. Where for example a 42 DE syrup is specified, the actual manufacturing tolerances are typically 41–44 DE. This is of no great consequence to the user of the syrup since there is very little practical difference in properties over this narrow range.

5.3 Degree of polymerisation (DP)

This term is widely used in the glucose industry as a shorthand means of describing the carbohydrate composition of a syrup. DP describes the number of glucose units in an individual component of a syrup. Thus DP1 = dextrose (1 unit), DP2 = maltose (2 units), DP3 = maltotriose (3 units) and so on and was basically introduced as a convenient means of describing these components.

5.4 Characterisation of glucose syrups

Where once DE was sufficient to describe the composition and hence properties of a glucose syrup it is now necessary to provide a more complete description owing to changes in manufacturing techniques.

The traditional method of acid hydrolysis produces a limited range of glucose syrups from starch and this remains a cost effective means of producing syrups for a range of industries, notably for confectionery. This basic method is however being replaced by acid–enzyme and enzyme–enzyme processes which extend the range of products available and hence widen the range of properties available to the food manufacturer. High fructose glucose syrups are a development of the enzyme–enzyme techniques whereby a high DE syrup is passed through a column of immobilised enzyme to convert a part of the dextrose to fructose. These processes are discussed in detail elsewhere in this book.

When specifying a glucose syrup the following parameters must be noted:

- Dextrose equivalent
- Carbohydrate composition – since it is possible to have several syrups of the same DE but with different composition and composition is important with regard to the syrup's properties
- Solids content
- Sulphur dioxide content

This will enable the syrup manufacturer to supply an exact product to meet the customer requirements.

The information provided in this chapter will relate the properties of glucose syrups broadly to their DE although as described above, this is not strictly speaking an exact relationship. Where obvious anomalies exist these will be described. The chapter is not intended as a comprehensive reference of data on glucose syrups but rather an indication of their properties and a guide to where further detailed information can be found.

5.5 Baumé

As a result of rather poor communications in the late 18th century several specific gravity scales were developed by different scientific communities. One such was produced in 1768 by Antoine Baumé, a French chemist.

The Baumé or Degrees Baumé (°Bé) scale was traditionally used by manufacturers of glucose syrups to describe the density or specific gravity of a syrup and thus indirectly to describe the syrup's solids content. It is determined using a hydrometer and as with refractive index, Baumé changes with DE and method of syrup manufacture at the same solids content.

In Europe the Baumé scale is becoming less widely used although in the USA it still finds application to describe glucose syrups, particularly with regard to specification requirements. Two scales are used, one for liquids lighter than water and one for liquids heavier than water. Since glucose syrups always have a specific gravity greater than water the scale

$$\text{Baumé} = 145 - \frac{145}{\text{true SG } 60°\text{F}/60°\text{F}}$$

where 145 represents the ratio of the total volume of water displaced by the hydrometer to the volume of water displaced by the unit scale length of the hydrometer stem. The results obtained using this relationship are often subtitled 'Modulus 145'. SG60°F/60°F = the specific gravity of glucose syrup at 60°F/specific gravity of water at 60°F.

Within the glucose industry Commercial Baumé is often used:

$$\text{Commercial Baumé} = \text{Observed Baumé at } 140°\text{F}/60°\text{F} + 1$$

The higher temperature of measurement reduces the viscosity of the syrup and allows the hydrometer to float more freely leading to a more rapid determination.

5.6 Bodying agent

Traditionally sucrose was used to provide both sweetness and mouthfeel (body) in soft drinks. More recently, glucose syrups have been used to complement the sugar to meet the consumer demand for reduced sweetness in these products although in some cases high fructose glucose syrups have been used to completely replace the sucrose. In this instance cost is the major consideration. Bodying effect or mouthfeel is related to the molecular weight and hence DE of a glucose syrup with predictably greatest effect, but least sweetness, with the lower DE products.

With the trends towards diet or reduced sugar drinks there would appear to be great potential for drinks made using a combination of low DE glucose syrup, to provide body, and intense sweetener, to provide sweetness. In canned fruit the bodying effect of the covering syrup is provided by glucose syrup and in biscuits the same material provides bite and chewiness.

5.7 Brix (degrees Brix)

This was developed and used almost exclusively by the sugar industry although like Baumé has found some application in the glucose sector.

Table 5.1 Relationship between Baumé, specific gravity and solids by weight

Commercial Baumé	Specific gravity	Solids content (% w/w)		
		30 DE	42 DE	52 DE
5	1.0367	10.47	10.53	10.58
10	1.0749	19.24	19.35	19.45
15	1.1161	28.06	28.24	28.39
20	1.1603	36.95	37.21	37.43
25	1.2083	45.92	46.27	46.57
30	1.2603	55.02	55.47	55.85
35	1.3169	64.25	64.83	65.30
40	1.3786	73.66	74.39	74.98
45	1.4463	83.36	84.25	84.98

Table 5.2 Relationship between Brix and Baumé (modulus 145)

Degrees Brix	Factor	Baumé
8.97	5	2.79
18.00	10	5.57
27.12	15	8.34
36.32	20	11.10
45.70	25	13.84
55.20	30	16.57
64.93	35	19.28
74.96	40	21.97
85.28	45	24.63

Instruments (hydrometers) are calibrated using sucrose and measurements are taken at 17.5°C. Refractometers are calibrated using sucrose and read directly in degrees Brix if measurements are made at 17.5°C. The literature provides a wealth of information concerning the relationships between Baumé, Brix, specific gravity and solids content of glucose syrups with general trends being given in Tables 5.1 and 5.2.

The factors in the central column of Table 5.2 can represent either degrees Brix or Baumé. If Brix then the corresponding Baumé value is given opposite in the right hand column and if Baumé the Brix value is given in the left hand column (values taken from Junk and Pancoast, 1973).

5.8 Boiling point elevation and freezing point depression

Both these colligative properties are dependent on the concentration and the molecular weight of the carbohydrate in solution and glucose syrups of different DE, and thus different average molecular weight, can be used to control these parameters. High DE syrups (low average molecular weight) will cause a greater elevation of boiling point and greater depression of freezing point than lower DE (high average molecular weight) syrups.

Table 5.3 shows the effect on boiling point elevation (Dziedzic and Kearsley, 1984) and Table 5.4 shows the effect on freezing point depression. Freezing point depression is calculated from average molecular weight of each syrup using the relationship:

$$\text{Freezing point depression} = \frac{1.86 \times \text{weight of solute} \times 100}{\text{average molecular weight} \times \text{weight of solvent}}$$

where $1.86 = K_f$ = molal depression constant for water (Glasstone and Lewis, 1965).

With respect to boiling point elevation, this effect is important in the

Table 5.3 Boiling point elevation of glucose syrups in relation to dry substance content (at normal atmospheric pressure)

	Boiling point (°C)					
	Dry substance (% w/w)					
DE	20	40	60	70	80	85
30	100.22	100.66	101.97	103.50	106.17	108.39
42	100.31	100.89	102.67	104.61	107.94	110.78
55	100.42	101.16	103.32	105.61	109.39	112.39
65	100.51	101.39	103.81	106.28	110.28	113.22
80	100.95	101.76	104.53	107.17	111.28	114.06
90	101.77	102.01	104.97	107.72	111.72	114.50

Table 5.4 Freezing point depression of glucose syrups in relation to dry substance content

	Freezing point depression (°C)				
	Dry substance (% w/w)				
DE	10	20	30	40	50
15	0.17	0.38	0.66	1.02	1.53
19	0.20	0.46	0.78	1.22	1.83
25	0.27	0.61	1.05	1.63	2.45
33	0.36	0.80	1.37	2.13	3.20
37	0.37	0.83	1.43	2.22	3.33
42	0.43	0.97	1.66	2.59	3.88
49	0.52	1.19	2.03	3.16	4.75
55	0.66	1.49	2.55	3.96	5.94
67	0.85	1.91	3.27	5.08	7.62
100	1.15	2.58	4.43	6.89	10.33
sucrose	0.60	1.36	2.33	3.63	5.44

production of hard boiled sweets where judicious selection of the glucose syrup to blend with sucrose can result in a lowering of the boiling temperature of the mix. This in turn can lead to reduced browning of the sweets due to overcooking and also, potentially, a decrease in energy requirements to produce the sweet.

Freezing point depression is important with respect to ice cream. For 'soft-scoop' products, which can be easily handled straight from the freezer, a large depression of freezing point is required and thus a high DE glucose syrup is used or even dextrose or fructose. Too low a freezing point may be undesirable however as this can cause handling and packing problems during manufacture. In such cases a lower DE syrup may be the better choice. In ice cream other factors such as sweetness and foam stabilising effects of the syrup are also important considerations.

5.9 Carbohydrate composition

The typical carbohydrate composition of a range of syrups is given in Figure 5.1. These figures illustrate the change in composition as molecular weight decreases (increase in DE) and also the effect of different hydrolysis techniques.

D.E. AND TYPE OF HYDROLYSIS	CARBOHYDRATE COMPOSITION				
	DP1	DP2	DP3	DP4	>DP4
15 DE ENZYME	4	4	5	5	82
30 DE ACID	10	9	10	8	63
37 DE ACID	14	12	10	9	55
37 DE ACID/ENZYME	8	30	12	4	46
37 DE ENZYME	5	35	14	3	43
55 DE ACID	31	18	13	10	28
63 DE ACID/ENZYME	37	32	11	4	16
63 DE ENZYME	33	46	3	2	16
70 DE ACID/ENZYME	43	32	7	5	13

Figure 5.1 Carbohydrate composition of glucose syrups.

5.10 Colour formation

The standard colour of glucose syrups when offered for sale is usually described as 'water white' or colourless. It is interesting to note that many food manufacturers specify water white and then proceed to add caramel or other colours to the syrup. In many applications water white can be replaced by less refined (and less costly) syrups. Water white syrups will not remain that way indefinitely as even fully refined syrups contain traces of proteins and amino acids, the reactions of which are described below. Colour formation (browning) in glucose syrups may occur as a result of two types of reaction:

(i) Maillard browning from reactions between reducing sugars and proteins or amino acids.
(ii) Caramelisation due to the effect of heat on carbohydrates.

5.10.1 Maillard reactions

The interactions of carbohydrates and proteins are well documented and represent a complex series of chemical reactions, with brown colour formation being a final stage in the series. The reactions are basically those between reducing sugars (aldoses and ketoses) and amino groups in proteins or free amino acids. Thus all glucose syrups and dextrose are potentially reactive as they contain free aldehyde groups whilst those containing fructose with free ketone groups tend to be even more reactive. It follows that the higher the DE of the syrup the more reactive it will be towards proteins/amino acids. This is well documented in the literature (e.g. Kearsley, 1978a). Amino acids are similarly more reactive than larger proteins.

In some food products browning reactions are desirable, e.g. in bread to give a brown crusty loaf and in toffees to give acceptable colour and flavour, but generally they are not desirable. Brown boiled sweets or condensed milk do not appeal to the majority of people.

There is little that can be done to control these free aldehyde or ketone groupings without changing the essential characteristics of the glucose syrups themselves and control of Maillard browning therefore centres around the control of free proteins or amino acids in the finished syrup or addition of chemical retardants to the syrup to slow the reactions to acceptable levels. Maillard browning is accelerated by heat and the brown colours in the syrups tend to develop during evaporation as the finished product is concentrated or during storage where the syrup must be kept warm (45°C–55°C typically) either to prevent crystallisation (high DE syrups) or to facilitate handling (reduce viscosity) of the product (low DE syrups).

Dextrose is a particularly potent reactant in this respect being much more reactive than maltose for example. It is reported that a high maltose syrup of the same DE as an acid converted syrup (with higher dextrose content) browns much less readily than the higher dextrose-containing product. Thus composition as well as DE should be considered when selecting a glucose syrup for an application where browning may be a potential problem.

The proteins in the syrups are present as impurities from the manufacturing process. Glucose syrups are mainly manufactured from maize or wheat starch both of which sources typically contain up to 0.3% protein and residual amino acids. These proteins are carried through to the glucose syrup in reduced concentrations, insoluble proteins being removed during the filtration stages of refining. The modern practice of fully refining glucose syrups also reduces protein/amino acid concentrations in the finished product. For example, in addition to removing inorganic impurities, demineralisation will also remove charged proteins/amino acids and carbon treatment, in addition to removing coloured compounds, also removes many of the precursors to the brown coloured compounds themselves. Syrups produced from other sources, such as potato or tapioca starch, suffer less in this respect owing to lower protein levels in the starting material.

Sulphur dioxide is commonly added to glucose syrups as a chemical retardant for browning reactions where levels up to 450 ppm are often used. It is not added as a preservative to glucose syrups although at high concentrations undoubtedly has some antimicrobial action. The sulphur dioxide forms a complex with the free aldehyde or ketone groups which reduces their further participation in reactions with proteins/amino acids. Fully refined products require lower concentrations of sulphur dioxide to give the same effect. It should be noted that some countries, e.g. Japan do not allow sulphur dioxide in foods and manufacturers producing foods for these markets usually specify very low levels of sulphur dioxide in the raw materials they purchase. Where sulphur dioxide is not added even greater care must be taken during handling and storage of the syrup.

Lowering of the pH of the product will also reduce Maillard browning as the initial reactions between reducing sugars and amino acids proceed more quickly at a higher pH. This has limited potential in glucose syrups which are typically produced to have a pH between 4.5–5.5.

Whilst it is possible to deliver a colourless glucose syrup to a food manufacturer the syrup producer has no control over how the user handles and stores the syrup and even more importantly with respect to browning, what other food ingredients are mixed with the syrup and what heat treatment this mixture receives. It is also important to remove impurities (e.g. amino acids as seen above) from glucose syrups during the refining stage as they can in some cases catalyse colour formation or may be precursors to coloured compounds. Demineralisation removes many charged species (minerals/amino acids) and carbon treatment removes any coloured compounds already formed in the syrup.

5.10.2 Caramelisation

Caramelisation of a carbohydrate occurs when it is heated excessively and particularly in the presence of an alkali. This forms the basis of a whole industry dedicated to the production of caramel colours for food use, e.g. in cola beverages, beers, gravy browning. The final coloured compounds produced during caramelisation are not dissimilar to those produced as a result of the Maillard reactions although proteins are not of course involved directly in caramelisation. The reaction develops round free aldehyde or ketone groups and caramelisation is therefore most likely in higher DE glucose syrups. Ketone groups are much more reactive than aldehyde in this respect and fructose-containing syrups will begin to caramelise at about 70°C. This is of course very important with respect to storage of these products where on the one hand high temperatures are needed to prevent dextrose crystallisation and low temperatures are required to prevent browning. Prolonged storage of fructose syrups is not recommended or normally carried out.

From a glucose syrup manufacturing viewpoint caramelisation reactions are less important than Maillard reactions although potentially they will occur if 'hot spots' are present in the process.

Whilst browning is essentially a problem associated with aqueous systems, it also occurs with some crystalline products. Crystalline fructose will for example start to brown above about 70°C whilst maltodextrins are stable even in the presence of proteins.

Since hydrogenation eliminates free aldehyde and ketone groups this process can be used to control both types of browning. Thus dextrose is converted to sorbitol, fructose to a mixture of sorbitol and mannitol and the other components of glucose syrups to the corresponding sugar alcohol.

5.11 Fermentability

This property, like many others, is desirable under some circumstances and undesirable under others. A soft drink manufacturer would not require his sweeteners to ferment yet a brewer or pharmaceutical manufacturer would require high fermentability. Fermentability of a glucose syrup may be defined as the percentage of its carbohydrate content, on a dry weight basis, that is fermented by yeast under controlled conditions (Corn Refiners Association, 1992).

The brewing industry in particular utilises high DE glucose syrups in the production of alcohol. Alternatively, high maltose syrups can be used for this purpose and these allow the mashing stage to be bypassed, with consequent increase in throughput for the brewery. Since the percentage of fermentable sugars decreases as DE decreases a means of controlling fermentation and alcohol production is available. In relation to glucose syrups, DE value is a useful indicator of the fermentability of the syrup. Thus a 21 DE product is about 21% fermentable, 43 DE 43% and 100 DE 100%.

The mono- and disaccharide components of a glucose syrup (dextrose and maltose) are rapidly fermented by both brewer's and baker's yeasts, the maltose after first being hydrolysed into its component parts. The trisaccharide component (maltotriose) is slowly fermented whilst higher saccharides are not fermented to any great extent. Both yeasts ferment maltose and maltotriose very slowly in the absence of glucose (Shieh *et al.*, 1973). Hence the approximate relationship between DE and fermentability.

High fructose glucose syrup is almost completely fermentable whilst the hydrogenated derivatives of glucose syrups, sorbitol, maltitol and hydrogenated glucose syrups themselves are non-fermentable. Figure 5.2 summarises this information. The fermentability of a glucose syrup can be controlled to a certain extent by the process of hydrogenation whereby the percentage of fermentable sugars in a syrup is reduced proportionally as the degree of hydrogenation is increased. Although glucose syrups of any DE can be hydrogenated to give a zero reaction to Fehling's solution it has been shown that with lower DE syrups some fermentability with yeast remains even when these products are completely reduced (Kearsley, 1978a).

In the gut hydrogenated glucose syrups are broken down to glucose and

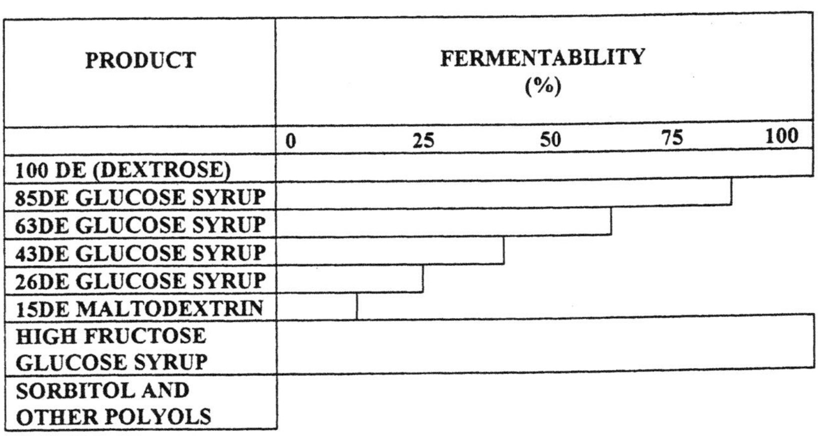

Figure 5.2 Fermentability of glucose syrups and related products.

sorbitol and predictably the laxative effects of the lower DE syrups are proportionally reduced compared with the higher DE syrups as they produce less sorbitol on hydrolysis. Whilst fermentation with yeast follows the pathways described above, other organisms used in fermentation will fully metabolise a range of glucose syrups and DE is a less important feature.

5.12 Foam development and stabilisation

If air is whipped into a glucose syrup or a food product containing a glucose syrup the rate at which the air bubbles pass out of that syrup will depend largely on the viscosity of the system. Thus lower DE syrups will retain air for longer periods than higher DE syrups as a result of their high viscosity and additionally they impart strength and some flexibility to the walls of the gas bubbles. They also thus slow the diffusion of air from the bubbles to the surrounding atmosphere and by maintaining bubble integrity slow diffusion from bubble to bubble.

In some foods air retention is a desirable attribute, e.g. ice cream and instant whips whilst in others it is not desirable, e.g. hard boiled sweets. Low DE products are thus used to trap air and stabilise aerated systems. For the same reason low DE products offer better possibilities for emulsion stabilisation than high DE products as they slow the coalescence of oil droplets. Conversely low molecular weight components (high DE syrups) improve the development of a foam by greatly increasing the flexibility of the walls of the air bubbles. A compromise must therefore be made between foam development and foam stabilisation when selecting a glucose syrup for this application.

5.13 Glaze formation

Glazing is widely used on baked goods to improve their appearance and give some surface protection, e.g. meat and fruit pies and cakes. A glaze must remain shiny throughout the life of the product, must not crystallise and critically must not absorb water to make the product turn ‘soggy’. Low DE glucose syrups or maltodextrins are typically used for this purpose.

5.14 Humectancy

A humectant is a substance whose moisture content changes very little even when the moisture content of its surroundings changes substantially. Humectancy is therefore different from hygroscopicity. A good humectant will for example prevent a cake from drying out when the cake is placed in a dry environment and conversely not make the cake 'soggy' when humidity increases.

Glucose syrups do not make particularly good humectants as their moisture contents change rapidly in response to humidity changes. The mid range syrups, 43–65 DE, come closest to providing a humectant solution for a food manufacturer but sorbitol syrup is undoubtedly the optimum humectant for food and other uses.

5.15 Hygroscopicity

Hygroscopicity of a carbohydrate is the ability to be able to absorb moisture from the surrounding atmosphere. This property may be desirable in some cases, e.g. cakes where the included carbohydrate attracts moisture to keep the cake moist and counters the drying out associated with such products. In most cases it is not desirable, however, a good example being boiled sweets. Control of moisture absorption is a role well suited to glucose syrups in foods.

Whether a glucose syrup or dried glucose syrup or other carbohydrate or a food product absorbs water from its surroundings depends on its equilibrium relative humidity (ERH). At a specific vapour pressure and temperature each product will neither gain nor lose moisture from its surroundings. This is the ERH of that product. Hygroscopicity is directly related to DE with the higher DE syrups being more hygroscopic. Additionally, the higher DE products, although starting as non-crystalline solids, rapidly liquefy as moisture is absorbed whilst the lower DE solids remain as such even at 100% humidity.

Products with a lower ERH than their surroundings will attract moisture, becoming sticky, and products with a higher ERH will tend to lose moisture and dry out. An increase in molecular concentration decreases the ERH and therefore a range of glucose syrups of different DE and hence molecular weight can be used to control the moisture absorbent properties of foods. Low DE products (high molecular weight = low molecular concentration) increase the ERH of a food and reduce its tendency to absorb moisture (but may lead to drying out) whilst higher DE products produce the opposite effect. It has also been reported that the lower DE syrups attain equilibrium more quickly than the higher DE products (Kearsley and Birch, 1975).

At the molecular level many studies concerning the hygroscopic nature of individual glucose polymers have been reported in the literature (Cleland and Fetzer, 1944; Donnelly *et al.*, 1973; Johnson and Srisuthep, 1975) as well as studies using commercial syrups (Mahdi and Hoover, 1965; Kearsley and Birch, 1975). Many conflicting results have been published but these appear largely due to experimental inconsistencies rather than any obvious differences in the syrups or glucose polymers themselves. It is known for example that the particle size of the dried glucose syrup, the method used to dry the syrup (probably resulting in different particle size ranges) and any impurities, e.g. minerals in the product, will affect its hygroscopic tendencies. The hygroscopic tendencies of individual oligomers have been reported and some interesting results obtained (Donnelly *et al.*, 1973). In this study it was apparent that the amount of water absorbed by different glucose oligomers did not follow the predicted path of DP1>DP2>DP3>DP4 etc. At 38°C and 90% relative humidity for example, the order was DP3>DP4=DP5=DP7>DP6>DP11>DP2.

In the case of minerals the effect appears most likely due to interactions between the carbohydrate and the inorganic material resulting in a loose complex formation and also due to inorganic material entering the helical structure of the higher molecular weight oligomers, again producing a complex product.

Chemical modification of a glucose syrup, for example by hydrogenation, leads to a significant reduction in moisture absorption at very high humidities (Kearsley, 1978a).

Table 5.5 gives typical values for moisture absorption of a range of glucose syrups after 14 days at 75% and 100% humidity and 25°C (Kearsley, 1978a). It is interesting to note that dextrose (100 DE) is less hygroscopic than even a 21 DE glucose syrup at 75% humidity. This is undoubtedly due to the fact that the anhydrous dextrose forms the monohydrate and that this is a relatively stable product. After 14 days the moisture absorbed by the dextrose exceeds that absorbed by the 65 DE product but is still less than that absorbed by the 85 DE material.

In systems containing mixtures of carbohydrates water absorption may not follow predicted patterns. For example a theoretical consideration of boiled sweets containing sucrose and glucose syrup might predict moisture absorption. In practice however, although a limited amount initially

Table 5.5 Moisture absorption by glucose syrup solids at 75% and 100% humidity

	Moisture absorbed (% w/w)	
DE	75% Humidity	100% Humidity
21	18	105
31	20	136
43	21	132
50	25	167
65	28	181
84	34	208
100	15	185

occurs, sucrose crystallisation at the sweet surface forms a microcrystalline layer which prevents further rapid moisture absorption.

5.16 Molecular weight and DE

The carbohydrate composition of a glucose syrup depends on the method used to produce that syrup, notably whether an acid or enzyme or combinations of these techniques are used to hydrolyse the starch. The properties of a syrup (or dried glucose syrup) in isolation or in a food product depend on its composition and in broad terms on its molecular weight. There is a logarithmic relationship between molecular weight and DE and molecular weight decreases as DE increases. It should be noted that syrups of the same DE but manufactured by different techniques will have slightly different molecular weights.

Table 5.6 illustrates the relationship between DE and average molecular weight calculated from freezing point depression measurements. The importance of demineralising a syrup before such determinations are made is obvious. Demineralisation is also of significance for accurate measurement of other properties of glucose type syrups such as boiling point elevation and osmotic pressure.

5.17 Nutritive solids

In resting subjects, blood glucose and serum insulin profiles are the same after consumption of any glucose syrup. Thus dextrose gives the same profiles as a 10 DE maltodextrin. This has been attributed to the fact that in such subjects absorption of dextrose through the gut wall is the limiting factor in the metabolism of a glucose syrup rather than the rate of hydrolysis in the gut.

Table 5.6 Average molecular weight versus DE

DE	Average molecular weight
15	1212
19	1019
25	760
33	580
37	558
42	479
49	392
55	313
67	244
100	180

In exercising subjects this may not be the case and athletes have been advised to take maltodextrin solution during periods of intense activity. It is thought that in these subjects the hydrolysis becomes the rate limiting step rather than the absorption and the glucose produced diffuses more slowly into the blood stream to provide a prolonged potential source of energy.

With babies, dextrose is known to induce diarrhoea in some instances and is thus avoided in baby foods in high concentrations. Starch, on the other hand, is difficult for the baby to digest. The solution therefore is to use a maltodextrin as the base for such products as these are relatively slowly digested by the baby and cause fewer digestive problems.

5.18 Osmotic pressure and water activity

Glucose syrups are essentially self-preserving and no preservatives need to be added to them to prevent microbial growth. This is due to the high dissolved solids content in the syrups effectively binding all the free water and thus generating a high osmotic pressure and low water activity. With no free water, microbial growth does not take place. It has been reported that some glucose oligomers possess antimicrobial activity which would undoubtedly contribute to their stability (Kondo *et al.*, 1975). Osmotic pressure is directly proportional to molecular concentration or average molecular weight and the lower the molecular weight the higher the osmotic pressure at the same concentration. Thus high DE syrups produce a correspondingly higher osmotic pressure than low DE syrups at the same concentration (higher molecular concentration) and are thus more effective against microbial spoilage. For glucose syrups below about 30 DE the solids content required to prevent microbial growth is so high as to prevent easy handling of the syrup and thus such products are commonly only available as spray dried powders.

Glucose syrups are not sterile and although no microbial development may occur in the syrup owing to the high osmotic pressure and low water activity the microorganisms in the syrup are not necessarily killed during storage although some reduction in numbers may occur. Any dilution of the syrup will reduce the osmotic pressure and increase the water activity and microbial growth may occur. Correct storage of glucose syrups is therefore essential to prevent condensation on the interior of the tank above the syrup as this could run back onto the surface of the syrup causing localised dilution. Trace heating of the headspace or air flow through the headspace is normally carried out to prevent condensation occurring.

Water activity is related both to the water content and the temperature of the particular syrup as well as its concentration and composition. Table 5.7 shows typical water activity values of glucose syrups. Note that bacteria do not develop when the water activity is below 0.9, moulds when below

Table 5.7 Water activity of glucose syrups at 20°C

Syrup	Solids content (% w/v)	Water activity
42 DE Acid glucose syrup	75	0.81
	80	0.77
	85	0.70
60 DE Acid-enzyme glucose syrup	75	0.78
	80	0.71
	85	0.64
90–94 DE glucose syrup	74	0.72
High fructose glucose syrup	70	0.76
Liquid sugar	67	0.85

0.8 and osmophilic yeasts when below 0.6. Thus all glucose syrups can be made safe from bacterial and mould spoilage but care should be taken to avoid osmophilic yeast contamination.

There is a linear relationship between DE and osmotic pressure at the same solids content and also between osmotic pressure and concentration at the same DE (Kearsley and Birch, 1978) and a method of DE determination has been developed based upon this principle (Kearsley, 1978b; Fitton, 1979). However, this method is only suitable for syrups which have been demineralised since small amounts of minerals can lead to errors in the determination owing to their effect on the osmotic pressure.

5.19 Prevention of crystallisation

Foods may contain high concentrations of individual sugars, e.g. jam and boiled sweets with a high sucrose content and where these high concentrations are found there is the risk of crystallisation of that sugar. Traditionally in jam and boiled sweet manufacture a combination of heat and acid was used to control the inversion of the sucrose to give a finished blend of sucrose, dextrose and fructose. Whilst potentially all may crystallise if present alone, providing no single sugar is present in high concentration, crystallisation in the system is avoided. Generally the greater the number of different components in a sugar mix the more difficult it is for one particular sugar to crystallise. Today, glucose syrups are used to 'dilute' the sugar and to provide the additional sugars to prevent crystallisation. They are ideally suited for this purpose being complex mixtures and such syrups are referred to as 'doctor syrups'. Additionally, low DE syrups containing high molecular weight glucose oligomers are able to form a matrix with low molecular weight sugars such as dextrose, fructose and sucrose and again prevent localised concentration buildup with attendant risk of crystallisation (Howling, 1992). Although glucose syrups are complex mixtures of dextrose and its polymers, only dextrose, maltose and maltotriose are crystalline with the higher saccharides being amorphous in nature.

5.20 Refractive index

The Refractive Index (RI) of a specific glucose syrup, as determined using a refractometer, is directly related to its dry solids content and is used as such for this purpose industrially to control evaporative processes. It is also used as a routine quality control acceptance technique for dry solids in a syrup as it is infinitely quicker to carry out a RI determination than to carry out the oven drying of a sample. It is important to note that RI is dependent on certain parameters of the syrup:

- RI increases as the solids content of the syrup increases.
- RI decreases as the DE of the syrup increases at the same solids content.
- RI decreases as the temperature of the syrup increases at the same solids content.

It is thus important to measure RI at a specific temperature (normally 20°C or 40°C) and by reference to appropriate tables, which take account of the effects of DE, obtain a value for the solids content of the syrup under test. This value will usually be slightly different from the true solids of the syrup owing to a combination of errors introduced from the above relationship. The difference will not usually be of commercial significance however and providing the purchaser and the manufacturer of the syrup use the same technique there will be few problems. The refractometer used for the test must of course be calibrated at the temperature of measurement.

The method by which the glucose syrup is produced, e.g. acid or acid/enzyme or enzyme/enzyme also influences the RI reading, all else being equal. Mineral content (ash) will also affect the RI of a syrup and corrections to the RI value may be necessary to take account of this as described in the literature.

Examples of the effects of temperature, solids content, DE and method of hydrolysis of the syrup on RI are shown in Table 5.8 and detailed information on the effects may be found in the literature (Wartman *et al.*, 1976). It is interesting to note in the literature (Okada and Nakakuki, 1992) that when the Refractive Index of different pure glucose oligomers from DP1 to DP10 was measured there was no difference between different products at the same concentration. This would tend to contradict to a certain extent the above findings.

Table 5.8 Typical refractive index values of glucose syrups

DE	Type of conversion	Dry solids (%w/w)	Refractive index	
			20°C	60°C
28	Acid	20	1.36534	1.35877
		40	1.40319	1.39592
42	Acid	20	1.36490	1.35834
		40	1.40222	1.39496
42	Enzyme/enzyme	20	1.36502	1.35846
		40	1.40244	1.39518
55	Acid	20	1.36463	1.35807
		40	1.40153	1.39427
63	Enzyme/enzyme	20	1.36444	1.35788
		40	1.40101	1.39376

5.21 Solubility

Glucose syrups are all very soluble in water, the limiting factor being the viscosity of the solution particularly with the lower DE materials. There is little difference in the overall solubility of different syrups but a significant difference in the rate at which the solid materials dissolve. This is of course particularly important for those products sold commercially as powders, typically below about 50 DE. Water temperature, rate of stirring and particle size are all important factors which will affect the rate of dissolution but these being equal, solubility rate is directly related to the DE of the material. Typical results are shown in Table 5.9 for products below 50 DE (Kearsley and Birch, 1975). Syrups below about 60 DE show little tendency to crystallise on storage even if storage temperatures fall. Above 60 DE dextrose may crystallise from solution if storage temperatures fall. The manufacturer and consumer of the glucose syrup therefore have the dilemma of storing at high temperature and risking browning or storage at lower temperatures and risking crystallisation. Obviously a compromise is reached and high DE syrups are typically stored at 45–55°C.

Table 5.9 Solubility rate for dried glucose syrups

Dried glucose syrup DE	Solubility rate (% min)
15	1.8
19	2.1
25	2.3
33	3.0
37	2.9
42	2.8
49	3.5

5.22 Specific heat

When designing equipment for producing glucose syrups, e.g. evaporators or for handling foodstuffs it is important to be able to calculate the heat input necessary to raise the temperature of the product. Table 5.10 shows typical values for 30, 42, 55 and 65 DE glucose syrups (Junk and Pancoast, 1973).

5.23 Specific rotation

Although specific rotation is a fundamental characteristic of a glucose syrup it has little relevance industrially except perhaps in the determination of complex mixture of sugars in foods (Pearson, 1976). When plane polarised light is passed through a glucose syrup solution the plane of polarisation is rotated as a result of the optical activity of the carbohydrates present in that syrup. Each component of the glucose syrup, dextrose, maltose, maltotriose etc., contributes to the overall optical rotation each having a different rotatory effect on the polarised light.

From the figure for optical rotation, specific rotation can be calculated and this figure is therefore related not only to the DE but also the composition of the particular syrup. Different glucose syrups of the same

Table 5.10 Specific heat of glucose syrups at 80% w/w solids at differing temperatures

	Specific heat (cal/g (°C))			
DE	26.7°C	48.9°C	71.1°C	93.3°C
30	137.2	141.5	145.8	148.8
42	139.0	143.5	148.0	151.8
55	140.7	145.8	150.6	154.6
65	142.5	147.8	152.6	156.6

Table 5.11 Specific rotation of glucose syrups

DE	Specific rotation (degrees)
15	+163
19	+168
25	+160
33	+153
42	+140
49	+130
55	+106
67	+97
100	+52.7

DE but produced by different means (acid, enzyme, acid–enzyme hydrolysis) will have a different specific rotation owing to their different carbohydrate composition. It is however approximately linearly related to DE (Birch, 1968).

The mineral content of the glucose syrup will affect the optical and hence specific rotation of a syrup possibly by inducing some conformational change in individual glucose polymers (Angyal and Pickles, 1972) or by forming helical complexes with the higher molecular weight oligomers. Table 5.11 gives typical values for the specific rotation of glucose syrups.

5.24 Sweetness

Sweet foods and drink are consumed far in excess of their value in relieving hunger and thirst. As a species we like and demand the sweet taste or sensation and a large sector of the food industry is concerned solely with the production of foodstuffs to meet this need. Traditionally sucrose and later glucose syrups were the sweeteners of choice for such foods and whilst these versatile raw materials are still widely used, artificial sweeteners are playing an ever increasing role in the foods we eat. The measurement of sweetness is not straightforward in the way that other parameters of glucose syrups are measured. There are no instrumental means of assessing sweetness and human tasters or assessors must be used with their associated errors. Trained taste panels can eliminate many of these and provided control of factors such as temperature of tasting, concentration and methodology is made, absolute and comparative values for sweetness can be obtained. As far as glucose syrups are concerned two general statements regarding sweetness can be made:

(i) sweetness increases as DE increases at the same concentration
(ii) sweetness increases as concentration increases at the same DE (up to saturation sweetness).

Additionally it is known that sweetness increases as tasting temperature increases and decreases with increasing acidity and lowering of pH. Additionally, the more viscous a solution the less sweet it tastes owing to impaired access of the sweetener to the taste receptor. All these factors are of course very important and vary widely in food systems although when assessing the sweetness of a product they are normally closely controlled. It is thus essential that not only is sweetness measured under closely controlled conditions to provide absolute values but also assessed in the actual finished product in which the sweetener is used.

Table 5.12 shows the sweetness values of a range of glucose syrups and related products using sucrose as the standard. For glucose syrups

Table 5.12 Sweetness of glucose syrups and related products

Carbohydrate	Relative sweetness
Sucrose	100
15 DE glucose syrup	14
25 DE glucose syrup	21
37 DE glucose syrup	27
43 DE glucose syrup	38
52 DE glucose syrup	43
64 DE glucose syrup	58
78 DE glucose syrup	63
86 DE glucose syrup	58
Dextrose	61
High fructose glucose syrup	90–95
Maltose	43
Sorbitol	63
Maltitol	75
Xylitol	90

threshold sweetness is inversely related to DE (logarithmically) and directly related to molecular weight (logarithmically).

It has been reported in the literature that components of glucose syrups with a molecular weight equal to or greater than maltotriose (DP3) possess little or no sweetness although some workers consider DP7 to be the cut-off point (Kimura and Nakakuki, 1990). This would imply that the sweetness of glucose syrups is mainly proportional to their glucose and maltose contents. Glucose syrups do however possess a sweetness greater than the sum of their glucose and maltose contents and synergistic reactions between components have been used to explain the effect.

Recent public interest in generally healthier lifestyles has resulted in changes in the diet, leading to the increased consumption of foods which are perceived as having beneficial physiological effects. One aspect of this has been the increased consumption of sugar-free confectionery. These products are considered non-cariogenic and contain fewer calories than the sweets they replace. In many cases they are also suitable for diabetics as they require no insulin for their metabolism.

Traditionally, sucrose and glucose syrups (with fats and texture modifiers as appropriate) have been used to produce confectionery. Sugar-free products are manufactured from sugar alcohols (hydrogenated derivatives of reducing sugars) the most commonly used being sorbitol, although others such as maltitol, isomalt and xylitol are widely used. The process of hydrogenation leads in some cases to quite dramatic increases in sweetness, for example with maltitol (Kearsley *et al.*, 1980).

High fructose corn syrup, produced by the isomerisation of a 90–95 DE glucose syrup (relative sweetness approximately 60) contains about 42%

fructose, 52% dextrose and 6% maltose. In the EU this is used mainly as a direct replacement for sugar in some soft drinks since it is approximately iso-sweet with sugar. The complete dominance of the soft drink market by HFGS and its wider application in Europe is limited not by its properties but by the production quota system in force in the EU.

Several fructose syrup suppliers now offer a range of lower fructose-containing syrups to extend their quota and although these are of reduced sweetness compared with the traditional 42% product they are still nevertheless used to replace sugar in many products owing to the demand for reduced sweetness foods.

In the USA where there are no restrictions on fructose production, the manufacture and use of fructose-containing syrups has developed into a second phase from the 42% fructose product (Bujake, 1986). A 55% fructose-containing syrup is now the industry norm and this has virtually eliminated the traditional soft drink sweetener, medium invert sugar, from the soft drink market in the USA as well as having successes in areas such as ice cream and baking. Additionally US producers also market 90% fructose syrups which are widely used in dairy products, salad dressings and canned fruits for example where they are used to provide sweetness but with reduced calories owing to the higher sweetness of the product compared with sucrose. The 55% fructose syrup has a sweetness of 99 and the 90% fructose a sweetness of 106 compared with sucrose (White, 1992).

5.25 Viscosity

The viscosity of a glucose syrup in relation to its solids content and temperature is an important factor when the product has to be pumped and stored, as heating may need to be provided for storage tanks and heavy duty pumps required to transfer the material. The viscosity of a glucose syrup is directly related to its moisture content and its molecular weight (and thus inversely related to DE). It is also related to the carbohydrate composition of the syrup and this effect is particularly significant when selecting a glucose syrup for confectionery manufacture. A 42 DE high maltose syrup for example has a lower viscosity than a 42 DE acid-converted syrup at the same solids content. Temperature is very important in relation to viscosity and viscosity decreases as temperature increases. Hydrogenation, predictably, has no apparent effect on the viscosity of a glucose syrup.

Table 5.13 shows typical values for viscosity of a range of glucose syrups at 75% solids content over a range of temperature (adapted from Corn Refiners Association, 1975).

Whilst the above data apply generally to glucose syrups different manufacturing techniques acid/acid–enzyme/enzyme–enzyme result in

Table 5.13 Relationship between viscosity, DE and temperature at 75% solids content

	Viscosity (cP)			
Temperature (°C)	30 DE	40 DE	60 DE	80 DE
15.5	70 000	20 200	5500	2000
26.7	19 000	6200	1700	700
37.8	5500	2200	650	300
48.9	2000	900	300	150
60.0	850	420	150	75
71.1	450	210	80	45

differences in viscosity at the same DE, all else being equal. For example, acid–enzyme and enzyme–enzyme 38 DE glucose syrups have a lower viscosity than acid 38 DE syrups. This is illustrated in Figure 5.3.

For bulk handling of glucose syrups the temperature of the product is critical both from viscosity and browning viewpoints and also with higher DE syrups to prevent dextrose crystallisation. Table 5.14 shows typical storage temperatures for a range of glucose syrups.

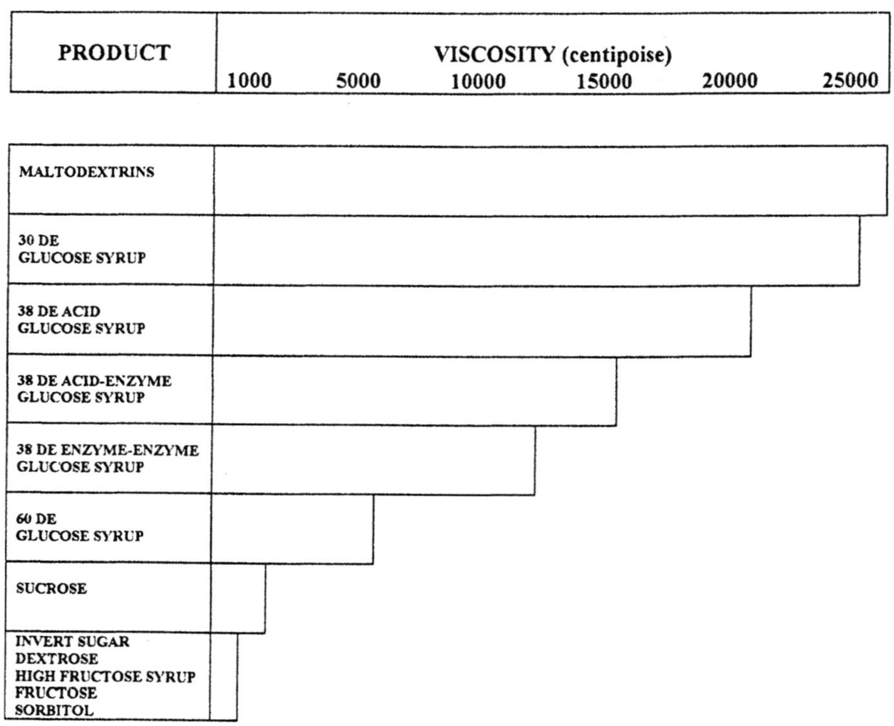

Figure 5.3 Comparative viscosities of glucose syrups and related carbohydrates at 70% w/w solids and 20°C.

Table 5.14 Storage temperatures of glucose syrups

Syrup	Solids content (% w/w)	Minimum storage temperature (°C)
38 DE Acid glucose syrup	75	40
	80	50
	85	60
38 DE Acid/enzyme glucose syrup	75	35
	80	45
	85	55
42 DE Acid glucose syrup	75	35
	80	45
	85	55
60 DE Acid/enzyme glucose syrup	75	20
	80	30
	85	45

Bodying effect is related to viscosity. This is of particular importance in soft drinks where it is often described as mouthfeel and is basically a description of the roundness or fullness of the syrup in the mouth. The higher the viscosity the greater the bodying effect.

FUNCTIONAL PROPERTY	TYPE OF GLUCOSE SYRUP HIGH MW/ LOW DE — LOW MW/HIGH DE
Bodying Agent	←————————
Browning Reaction	————————→
Cohesiveness	←————————
Emulsion Stabiliser	←————————
Fermentability	————————→
Flavour Enhancement	————————→
Flavour Transfer Medium	————————→
Foam Stabiliser	←————————
Freezing Point Depression	————————→
Glazing Agent	←———————→
Humectancy	←———————→
Hygroscopicity	————————→
Nutritive Solids	←———————→
Osmotic Pressure	————————→
Preservative Effect	————————→
Prevention of Sugar crystallisation	←————————
Prevention of Coarse Ice Crystals during Freezing	←————————
Sweetness	————————→
Viscosity	←————————

Figure 5.4 Properties and functional uses of glucose syrups.

5.26 Conclusions

Glucose syrups possess a wide range of properties which are largely due to the diverse composition of these products. These are summarised in relation to DE in Figure 5.4. As such they can be used to advantage in virtually all foods to control and contribute to the essential characteristics of that food. Their use in 'native' and 'modified' forms is set to increase in the future not only in developing countries but also in developed countries where they are able to provide the functional variety demanded by manufacturers of their ingredients. It is likely in the future that they will be sold ready blended with other carbohydrates to facilitate their use by large food and drink manufacturers and tailor made for specific nutritional needs.

References

Angyal, S.J. and Pickles, V.A. (1972) *Aust. J. Chem.* **25**, 1695.

Birch, G.G. (1968) *J. Food Technol.* **3**, 159.

Bujake, J.E. (1986) In *Alternative Sweeteners* (eds. L. Nabors and R.C. Gelardi), Marcel Decker.

Cleland, J.E. and Fetzer, W.R. (1944) *Ind. Eng. Chem.* **36**, 552.

Corn Refiners Association (1975) *Critical Data Tables*, Corn Refiners Association Inc., 1100 Connecticut Ave., N.W. Washington D.C., 20036.

Corn Refiners Association (1992) *Standard Analytical Methods*, Corn Refiners Association Inc., 1100 Connecticut Avenue, N.W. Washington D.C., 20036.

Donnelly, B.J., Fruin, J.C. and Scallet, B.L. (1973) *Cereal Chem.* **50**, 512.

Dziedzic, S.Z. and Kearsley, M.W. (eds) (1984) *Glucose Syrups: Science and Technology*, Elsevier Applied Science Publishers.

Fitton, M.G. (1979) *Stärke* **31**, 381.

Glasstone, S. and Lewis, D. (1965) *Elements of Physical Chemistry*, 2nd ed., Macmillan.

Howling, D. (1992) In *Starch Hydrolysis Products* (eds. F.W. Schenke and R.E. Hebeda), VCH.

Johnson, J.A. and Srisuthep, R. (1975) *Cereal Chem.* **52**, 70.

Junk, W.R. and Pancoast, H.M. (1973) *Handbook of Sugars*, Avi Publishing.

Kearsley, M.W. (1978a) *J. Food Technol.* **13**, 339.

Kearsley, M.W. (1978b) *J. Assoc. Public Anal.* **16**, 85.

Kearsley, M.W. and Birch, G.G. (1975) *J. Food Technol.* **10**, 625.

Kearsley, M.W. and Birch, G.G. (1978) *Stärke* **29**, 425.

Kearsley, M.W. Dziedzic, S.Z., Birch, G.G. and Smith, P.D. (1980) *Stärke* **32**, 244.

Kimura, K. and Nakakuki, T. (1990) *Stärke* **42**, 151.

Kondo, H., Honke, T., Hasegawa, R., Shimoda, T. and Nakamura, S. (1975) *J. Antibiot.* **28**, 157.

Mahdi, A.A. and Hoover, W.J. (1965) *Food Technol. Campaign* **19**, 123.

Okada, M. and Nakakuki, T. (1992) In *Starch Hydrolysis Products* (eds. F.W. Schenke and R.E. Hebeda), VCH.

Pearson, D. (1976) *The Chemical Analysis of Foods*, 7th edn, Churchill Livingstone.

Shieh, K.K., Donnelly, B.J. and Scallet, B.L. (1973) *Cereal Chem.* **50**, 169.

Wartman, A.M., Hagberg, C. and Ellason, M.A. (1976) *J. Chem. Eng. Data* **21**, 459.

White, J.S. (1992) In *Starch Hydrolysis Products*, (eds. F.W. Schenke and R.E. Hebeda), VCH.

6 Polyols from starch

Y. LE BOT and P.A. GOUY

6.1 Introduction – What are polyols?

Polyols, or polyalcohols, are a well-known family of products widely used throughout the world. In the European Union (EU) the family of polyols includes six products; isomalt, lactitol, maltitol, mannitol, sorbitol and xylitol. In the USA these products are described as sugar alcohols chemically defined as saccharide derivatives in which a ketone or aldehyde group is replaced by a hydroxyl group, and whose use is listed by the Food and Drug Administration or is generally recognized as safe.

Numerous fruits contain large amounts of polyols which contribute significantly to their sweetness (Sicard and Le Bot, 1994) (see Table 6.1). These fruits and vegetables represent the largest proportion of polyols in human consumption. The second source comes from the use of polyols obtained by hydrogenation of sugars. The main part of this consumption occurs through sugarless, safe for the teeth, confectionery.

Whereas isomalt is obtained from sucrose and lactitol from lactose, the other polyols can be obtained from sugars derived from starch. A biochemical conversion followed by purification leads to the production of sugars which are then hydrogenated (Le Bot, 1994). In this way glucose leads to sorbitol, fructose to a mixture of mannitol and sorbitol and maltose to maltitol (see Table 6.2).

Polyols are utilized either because of their sweetness or their technological functional properties. They are bulk sweeteners whose sweetness is generally lower than that of sucrose. Their sweetness is of particular interest since polyols, unlike sugars, do not contribute to dental caries formation. Advantage can also be taken of their reduced energy content (their calorific value is nearly half that of traditional sugars). They can therefore play a part in the formulation of 'light' foods. They are useful sugar substitutes providing manufacturers with the means to meet the needs of the consumers, eager to enjoy safe and dietetic products whilst not being prepared to compromise on quality.

Beyond their sweetness function, in chocolate and confectionery for instance, polyols are used for a wide variety of technological reasons in various food applications. They have, for instance, emulsifying, stabilizing, humectancy, thickening and cryoprotecting properties which are advantageous to the food industry.

Table 6.1 Natural occurrence of polyols

Polyol	Source of polyol	Content of polyol	
D-Sorbitol Heaton *et al.* (1980)	Apples	2.6–9.2	g D-Sorbitol/l of juice
	Pears	11.0–26.4	g D-Sorbitol/l of juice
	Cherries	14.7–21.3	g D-Sorbitol/l of juice
	Sour cherries	13.1–29.8	g D-Sorbitol/l of juice
	Plums	1.8–13.5	g D-Sorbitol/l of juice
	Rowan (*S. aucuparia*)	85	g D-Sorbitol/l of juice
D-Mannitol Sicard and Leroy (1983)	*Laminaria* spp.	10%	of dry weight
	Lactarius	15–20%	of dry weight
	Agaricus	15–20%	of dry weight
Xylitol Washuttl *et al.* (1973)	Yellow plums	0.93%	of dry weight
	Strawberries	0.36%	of dry weight
	Cauliflower	0.30%	of dry weight

Table 6.2 Origins of polyols

Beet or cane	Sucrose	Isomaltulose	α-D-Glucopyranosyl-1.6 sorbitol, α-D-glucopyranosyl-1,6 mannitol, isomalt
Corn, wheat	Starch	Maltose, corn syrup, maltose syrup, glucose syrup	Maltitol, maltitol syrup
Corn, wheat	Starch	Fructose	Mannitol/sorbitol
Corn, wheat	Starch	Glucose, dextrose, glucose syrup	Sorbitol, sorbitol syrup
Corn, wheat, birch bark	Starch hemicellulose	Xylose	Xylitol

6.2 Polyols and their legal status

6.2.1 Polyols – legal status in the EU

In the EU polyols are considered as food additives and are given the following code numbers:

- Sorbitol E420
- Mannitol E421
- Isomalt E953
- Maltitol E965
- Lactitol E966
- Xylitol E967

They are listed in two directives. The *Sweeteners Directive* and the *Directive on Food Additives other than Colours and Sweeteners.*

The *Sweeteners Directive* was adopted in June 1994, and published in the *Official Journal of the European Communities* (*OJEC*) on 10 September 1994. The polyols are listed in the directive's annex. They are authorized for use at *quantum satis* in 16 categories of foodstuffs that are either 'energy reduced' or 'with no added sugar'. Energy reduced is defined in the directive as 'with an energy value reduced by at least 30% compared with the original foodstuff or a similar product'. With no added sugar means "without any added mono- or disaccharides or any other foodstuff used for its sweetening properties". In addition, they are permitted in confectionery with no added sugar (but not energy reduced), in sauces, mustard, products intended for particular nutritional uses and solid food supplements/dietary integrators.

Use of the polyols for non-sweetening or 'technological' purposes, will be regulated at EU level by the *Directive on Food Additives other than Colours and Sweeteners*, which was adopted on 15 December 1994. Publication of this directive in the *OJEC* is expected early in 1995. The polyols are listed in annex IV (Other Permitted Additives) and in annex V (Permitted Carriers and Carrier Solvents). In annex IV, the condition of use is *quantum satis* in foodstuffs in general apart from drinks and unprocessed foods, with the exception of frozen and deep frozen fish, crustaceans, molluscs and cephalopods, where they are permitted. They are also authorized for use in liqueurs. There is no restriction on their use as carriers or carrier solvents.

6.2.2 Polyols – legal status in the USA

As mentioned previously a working definition has been accepted for sugar alcohols in the USA. Polyols listed in the European sweeteners directives and meeting this definition have to be considered individually: sorbitol is considered as a direct food substance affirmed as GRAS (generally recognized as safe). Petitions have been filed for GRAS affirmation for maltitol, maltitol syrup, isomalt and lactitol. Mannitol is a food additive permitted in food on an interim basis. Xylitol is a food additive which may be used in food for special dietary uses. It is nowadays also used in other kinds of food products (chewing gum).

6.2.3 Polyols – legal status in Japan

Maltitol and maltitol syrups are approved as 'natural food' and 'chemically synthesized food'. Sorbitol, on the other hand, is approved as a 'chemically synthesized food additive' without any restriction. Mannitol is also considered as a 'chemically synthesized food additive' and was until 1992 limited to 4%. Recently a draft regulation has been published allowing the

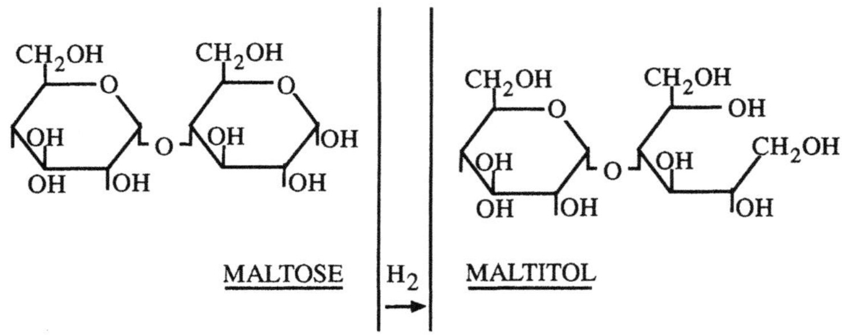

Figure 6.1 Hydrogenation of maltose to maltitol.

use of mannitol up to 20% in chewing gums and 40% in candies. Xylitol is not yet approved.

6.3 General points in the method of manufacture

6.3.1 Maltitol and maltitol syrup E965

Maltitol is produced commercially by catalytic hydrogenation of D-maltose (Figure 6.1). D-Maltose is a naturally occurring disaccharide and is a major constituent of malt wort. Industrially maltose is obtained by the enzymic hydrolysis of highly purified starch. Maltitol is a hydrogenated disaccharide having the empirical formula $C_{12}H_{24}O_{11}$.

Industrially in the manufacture of maltitol, the raw material used for the hydrogenation is a high maltose glucose syrup (Figure 6.2). High maltose glucose syrups are products with a high content of maltose (>50%) produced by an accurately controlled hydrolysis of purified starch using appropriate enzymes. In addition to maltose itself, these syrups contain higher molecular weight saccharide species (oligo- and polysaccharides). Maltitol syrups with maltitol contents ranging from 50–55% (Lycasin®) to 85% and higher are commercially available. A very pure maltitol (chemical purity >99%) is produced by a patented process (Devos and Gouy, 1985).

6.3.2 Mannitol E421

Mannitol takes its name from manna, a sweet exudate from the manna ash (*Fraxinus ornus*), of which it is the major constituent. Mannitol also occurs in olives, figs and larches, in certain species of edible fungi and in seaweed (*Laminaria* species).

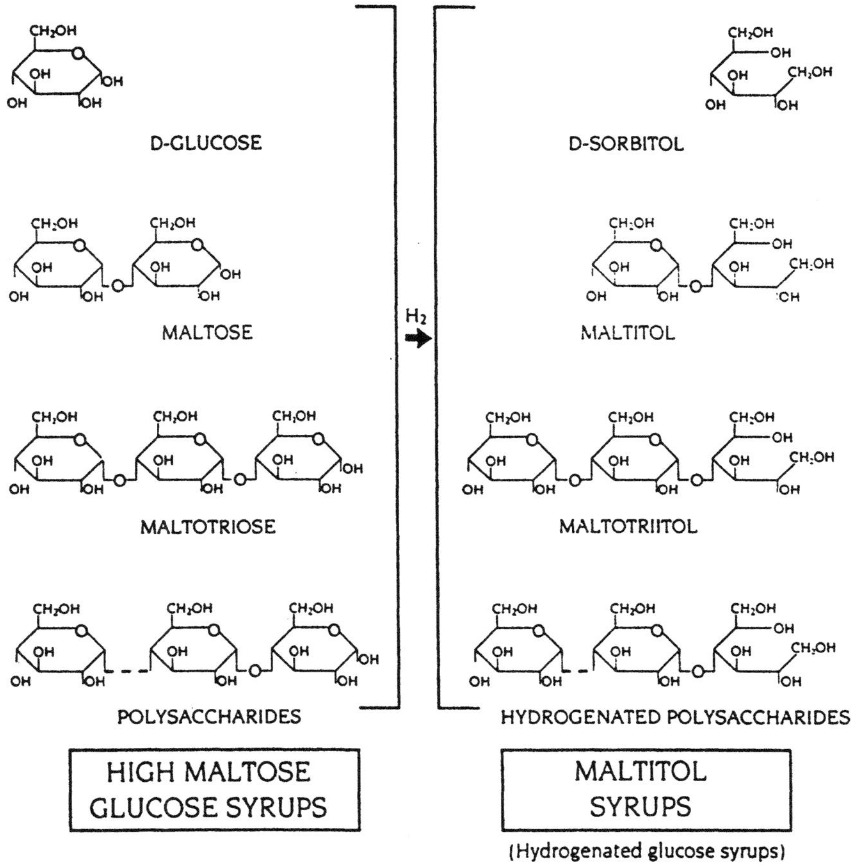

Figure 6.2 High maltose glucose syrups hydrogenated to form maltitol syrups.

D-Mannitol was originally obtained industrially by an extraction process; today it is produced by catalytic hydrogenation of D-fructose (Figure 6.3). Fructose is itself produced from invert sugar or starch. Through hydrogenation, a molecule of hydrogen is added to the ketone group of fructose, thus reducing it to an hydroxyl group. Mannitol is described as a hexitol, having six alcohol groups.

6.3.3 Sorbitol E420

Sorbitol takes its name from the sorbus tree (*Sorbus aucuparia*, the rowan or mountain ash) from the berries of which it was first extracted in 1872. It

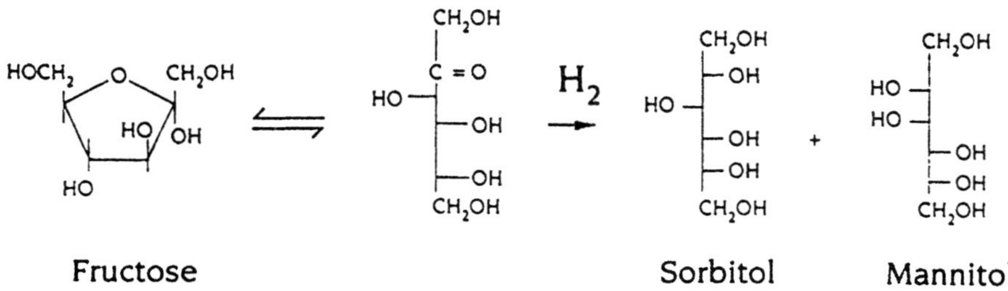

Figure 6.3 Hydrogenation of fructose to sorbitol and mannitol.

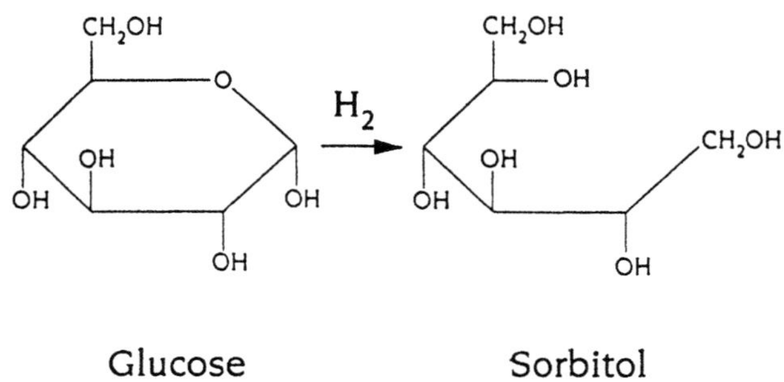

Figure 6.4 Hydrogenation of glucose to sorbitol.

occurs widely in nature in many fruits and vegetables, and in fact is also present in carbohydrate metabolism in man.

D-Sorbitol is produced commercially by catalytic hydrogenation of D-glucose (dextrose) (Figure 6.4). D-Glucose is also abundant in nature. It is also obtained by enzymatic hydrolysis of highly purified starch. The hydrogenation of dextrose involves the addition of a molecule of hydrogen to the aldehyde group (which exists as a hemiacetal in the pyranose ring form of D-glucose). The aldehyde group is thus reduced to an hydroxyl, or alcohol group. Sorbitol, like mannitol, is therefore a hexitol containing six alcohol groups.

6.3.4 *Xylitol E967*

Xylitol is present in many fruits and vegetables (e.g. greengages, strawberries, raspberries and cauliflowers) and is also naturally present in the human body. Xylitol is produced commercially by catalytic hydrogenation of the five carbon sugar xylose (Figure 6.5). This in turn is derived from xylan extracted from the hemicelluloses occurring in corn cobs,

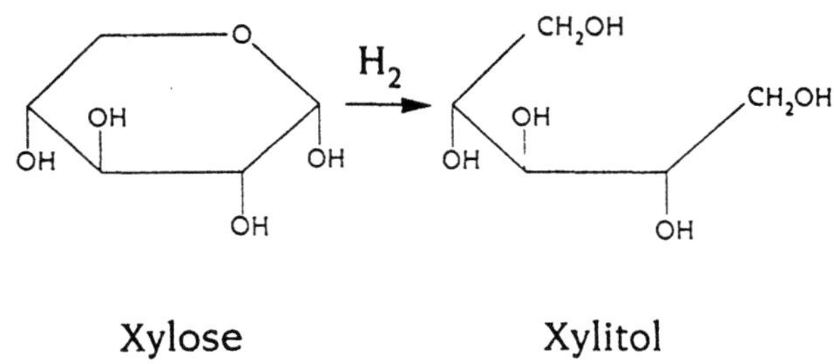

Figure 6.5 Hydrogenation of xylose to xylitol.

almond shells and birch bark. As in the production of sorbitol, a molecule of hydrogen is added to the aldehyde group of xylose, thus reducing it to an hydroxyl group. Xylitol, having five alcohol groups is a pentitol.

6.4 Physical and chemical properties of polyols

Sorbitol, mannitol, xylitol and maltitol are marketed as white crystalline powders, with a range of particle sizes and crystalline forms. Sorbitol and maltitol are also available as aqueous solutions.

6.4.1 Solubility

Most of the commercial syrups contain significant proportions of additional polyol species which influence the solubility and have the effect of controlling crystallization (Figure 6.6).

6.4.2 Viscosity of solutions

The viscosity of a syrup depends on the molecular weight of the solute. Monosaccharides (e.g. dextrose) and related polyols (e.g. sorbitol) give low viscosity solutions. Disaccharides (e.g. maltose) and related polyols (e.g. maltitol) give slightly higher viscosities (Figure 6.7). The presence of higher molecular weight polysaccharides leads to a greatly increased viscosity. Those commercial syrups with the highest content of higher molecular weight species have the highest viscosities.

6.4.3 Hygroscopicity

There is no relationship between the hygroscopicity of a polyol in solution and that of the same product in the solid form. Sorbitol is an excellent

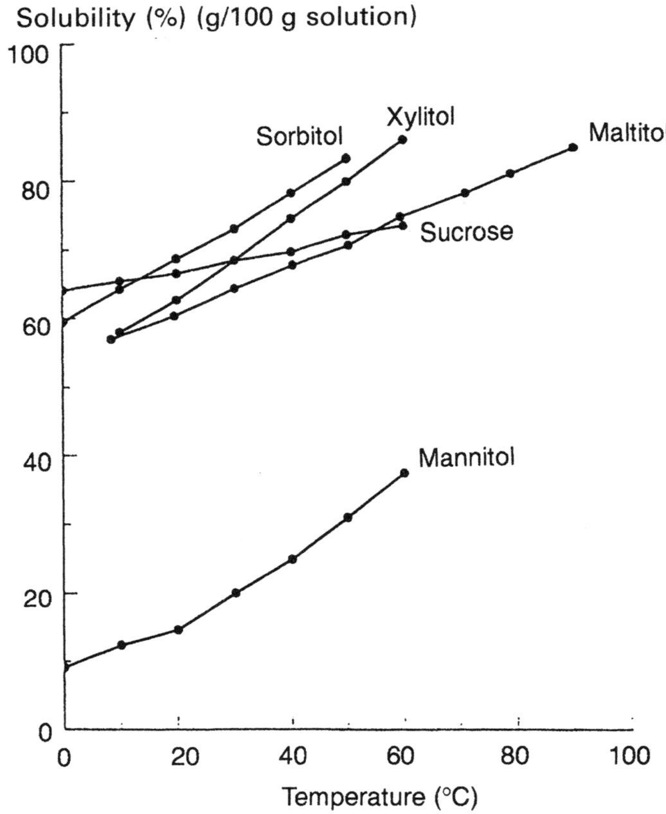

Figure 6.6 Solubility of pure polyols compared with sucrose.

humectant in solution owing to its high solubility. However, crystalline sorbitol is not very hygroscopic, particularly in the stable γ-form. Mannitol powder has very low hygroscopicity, taking up moisture only in excess of 98% relative humidity (RH). At 20°C pure crystalline maltitol (Maltisorb®) is less hygroscopic than sucrose (Figure 6.8).

6.4.4 *Humectancy*

The role of humectants is to limit the exchange of water to and from water-containing products. Depending on the application, in food or in cosmetics, sorbitol (Neosorb®) and liquid polyols used at concentrations of 3 to 60%, stabilize the moisture content, improve storage and give good consistency and plasticity.

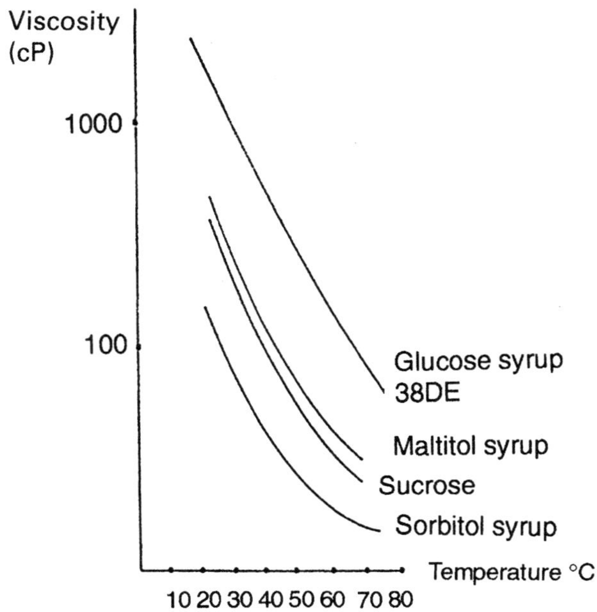

Figure 6.7 Viscosity of aqueous solutions at 70% DS (dry substance).

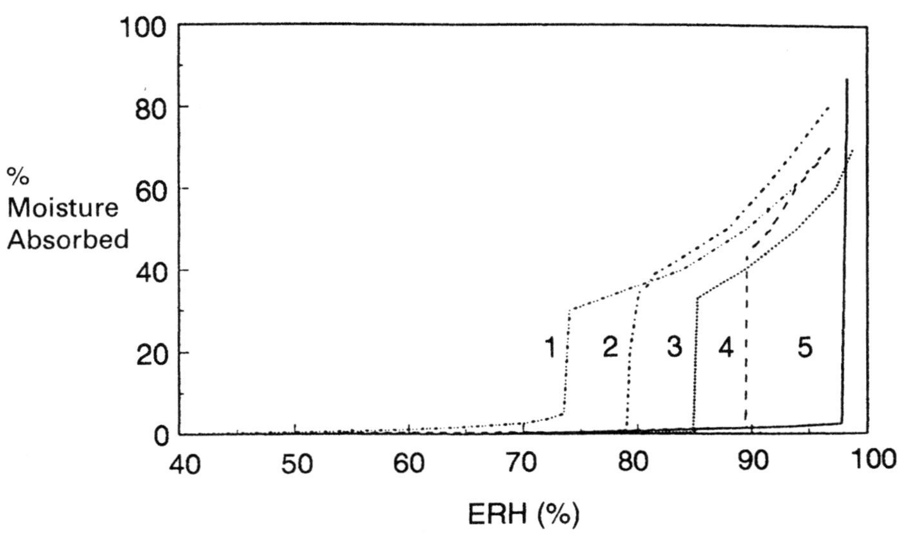

Figure 6.8 Hygroscopicity of crystalline powders at 20°C. 1, Sorbitol; 2, xylitol; 3, sucrose; 4, maltitol; 5, mannitol.

6.4.5 Compressibility

Sorbitol is directly compressible because of the dentritic structure of its crystals. Special grades of other polyols for direct compression, notably mannitol, have also been developed (Serpelloni, 1985).

6.4.6 Chemical reactivity

Polyols contain primary and secondary hydroxyl groups which are less reactive than the aldehyde and ketone groups of sugars. The polyols are all very stable to heat and melt without decomposition. Polyols do not react with amino acids. With certain grades some Maillard reaction browning, resulting from residual reducing sugars, may occur during the manufacturing process. Hydrolysis of maltitol and of the other higher molecular weight components of maltitol syrups is very slow under the mild conditions normally encountered in their use.

6.4.7 Melting point

Values of melting points are sorbitol (γ-form) 99°C, maltitol 147°C, mannitol 165°C and xylitol 94°C.

6.4.8 Boiling point, freezing point and osmotic pressure

These colligative properties, as with their patent sugars, are dependent on the molecular weight and concentration of solution.

6.4.9 Refractive index

The refractive index of polyol solutions increases linearly with concentration and is a method commonly used to specify the concentration of a given syrup.

6.5 Organoleptic properties of polyols

6.5.1 Sweetness

Sweetness is subjective and depends on the individuals who are doing the tasting and their physical condition. A number of other variables such as concentration, temperature, pH and the presence of other sweeteners and food ingredients can influence apparent sweetness significantly. Xylitol is the sweetest polyol as it is equivalent to sucrose. Pure crystalline maltitol

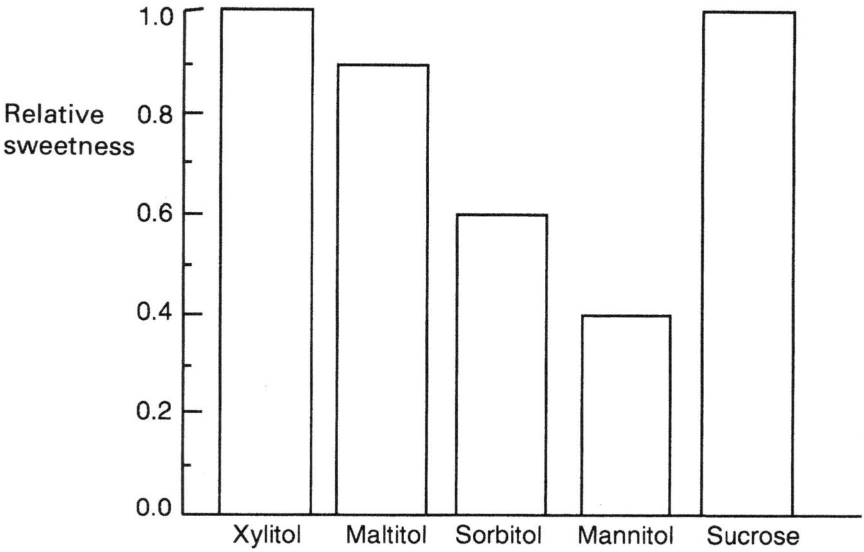

Figure 6.9 Relative sweetness of polyols.

has about 90% of the sweetness of sucrose. The sweetness of maltitol syrups depends on their maltitol content (Figure 6.9).

6.5.2 Cooling effect

A characteristic property of all polyols is their negative heat of solution which gives a cooling sensation in the mouth. This cooling effect (Figure 6.10) depends upon:

- the heat of solution
- the solubility of the polyol at mouth temperature (37°C)
- the particle size (the finer the product, the more quickly it dissolves and the cooler and sweeter it seems).

Sorbitol and xylitol provide the greatest cooling effect in the mouth. Mannitol, although it has a higher heat of solution than sorbitol, does not have a very strong cooling effect owing to its low solubility (Table 6.3).

6.5.3 Other taste characteristics and mouth sensations

All the polyols have a pleasant, clean and neutral taste in solution. Their body, mouthfeel and taste profiles can be compared with those of sucrose.

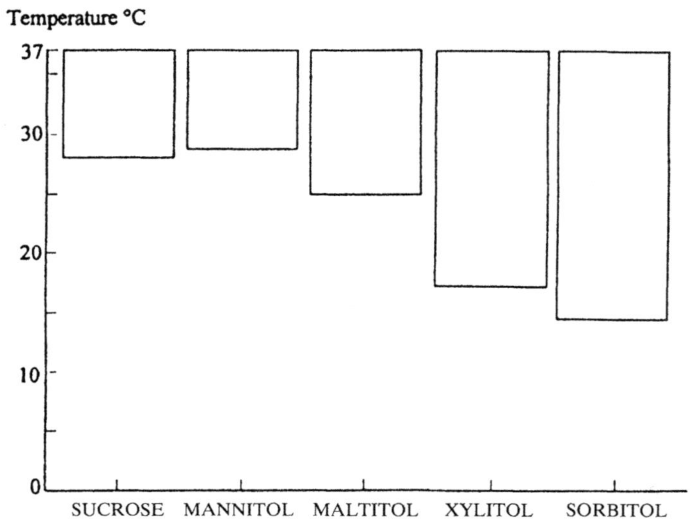

Figure 6.10 Cooling effect of polyols.

Table 6.3 Heat of solution and solubility of polyols

	Xylitol	Mannitol	Sorbitol	Maltitol	Sucrose
Heat of solution (cal/g)	–34.8	–30.0	–28.0	–16.3	–4.3
Solubility at 37°C (g/100 ml water)	256	32	334	201	229

6.6 Nutritional aspects of polyols

6.6.1 *Metabolism and calorific value*

6.6.1.1 The official EU calorific value. The EU has agreed that in calculating the energy value of foods the calorific value of all polyols shall be 2.4 kcal g^{-1}. This is to be compared with a value of 4 kcal g^{-1} for sugars and other carbohydrates.

6.6.1.2 Intestinal metabolism. Sugars are completely absorbed in the intestine through an active transport mechanism for glucose. In their subsequent metabolism they contribute a calorific value of 4 kcal g^{-1}. With polyols the majority is not absorbed and what reaches the colon is fermented by the microflora to give volatile fatty acids. This fermentation represents a calorific value of about 2 kcal g^{-1}.

There are some minor differences between various polyols as can be

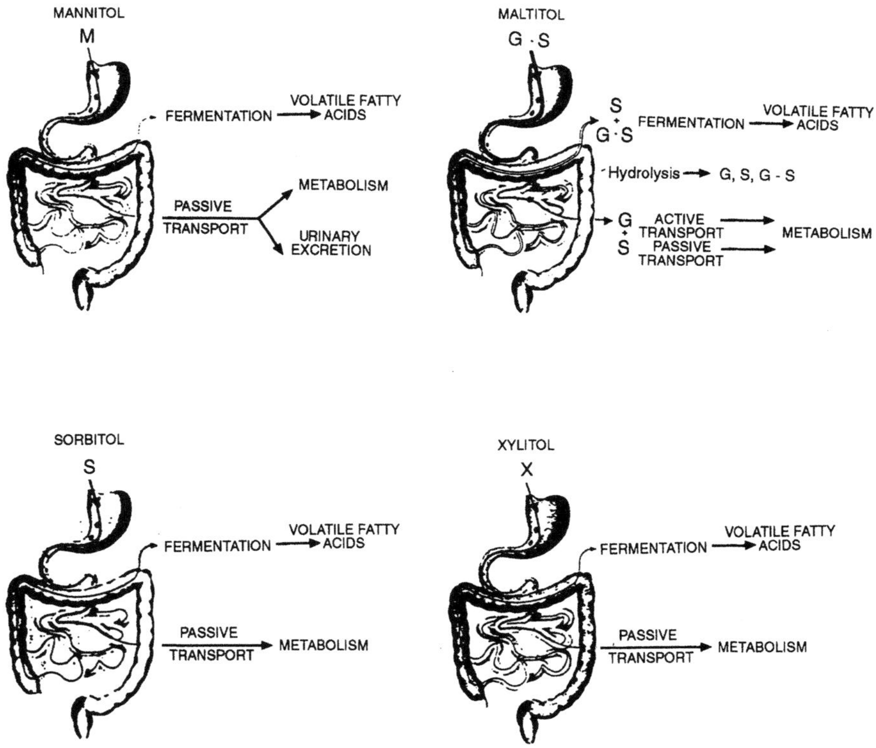

Figure 6.11 Intestinal metabolism of polyols.

seen from Figure 6.11. Only a small proportion of ingested maltitol, mannitol, sorbitol and xylitol is absorbed through the small intestine wall by a slow passive diffusion. This fraction, therefore, only contributes a small part to the calorific value of polyols. Part of the mannitol which is absorbed is eliminated by urinary excretion. For maltitol to cross the intestine wall it must first be split by hydrolysis which very slowly generates glucose and sorbitol.

6.6.1.3 Polyols and diabetic diet. The enzyme sorbitol dehydrogenase (found mainly in the liver, the prostate and the kidneys) enables D-sorbitol to enter the normal metabolic pathway for sugars at the level of D-fructose. The oxidation of D-sorbitol and the subsequent oxidation of D-fructose are hepatic reactions which allow these compounds to enter the glycogenolytic pathways through an insulin-independent mechanism. The enzymes involved, D-sorbitol (or L-iditol) dehydrogenase, fructokinase and hepatic aldolase, are not insulin dependent in contrast to glucokinase which converts D-glucose into Glucose-6-P. This explains why D-sorbitol may replace D-glucose and sucrose in the diet for diabetics.

In contrast to D-glucose, there is no active transport mechanism for absorption of D-sorbitol by the body. Passive absorption has been estimated at 10–20 g h^{-1}, but the value varies widely according to the physiological state of the individual. Sorbitol has the advantage of not giving rise to elevated blood glucose levels when eaten. Glucose may be generated during the metabolism of sorbitol although, owing to the delay in its production, it does not lead to any appreciable hyperglycaemia.

Mannitol and xylitol also induce very low glycaemia and are therefore suitable for diabetics.

Maltitol and maltitol syrups yield glucose and sorbitol when hydrolysed, thus their metabolism relates to that of these materials. The oligosaccharide content of maltitol syrups is hydrolysed in the intestine into glucose and maltitol as a first step, and then the maltitol is slowly hydrolysed as a second step. Owing to the slow rate of the maltitol hydrolysis, the glycaemic index is low and there is a low insulin response.

6.6.2 Polyols – dental aspects

Dental caries is caused by the fermentation of sugars by the buccal flora which in turn leads to acid production and decalcification. Many studies and official reports have been published advising the use of sugar substitutes. Polyols are not fermented by oral bacteria and are therefore not acidogenic or cariogenic. Replacement of the fermentable sugars by polyols is thus a sound option, especially in confectionery and liquid medicines.

The San Antonio Integration of Methods Working Group (A.D.A., 1986) reached the conclusion that some food could be categorized as having no cariogenic potential. The group proposed that measurements of the pH of dental plaque (pH telemetry) should be an approach capable of identifying food possessing no cariogenic potential. Specifically, food that results in pH profiles statistically equivalent to those generated by sorbitol would be deemed as possessing no cariogenic potential. It is relevant to note that sorbitol was chosen as the control.

Professor Muhlemann at the Zurich Dental Institute developed a test to measure the acidogenic potential of food *in vivo*. In Switzerland it is now possible legally to claim that a product is ‘safe for teeth’ only if it has undergone tests to prove that the plaque pH does not fall below 5.7 within 30 min of consuming the product (Figure 6.12). Sorbitol, mannitol, xylitol, maltitol, maltitol syrup and Lycasin have all passed the test (Imfeld, 1983).

Having passed this test it is possible to use these polyols to manufacture sugar-free confectionery to make use of the logo of the toothfriendly sweets association, Sympadent (Figure 6.13).

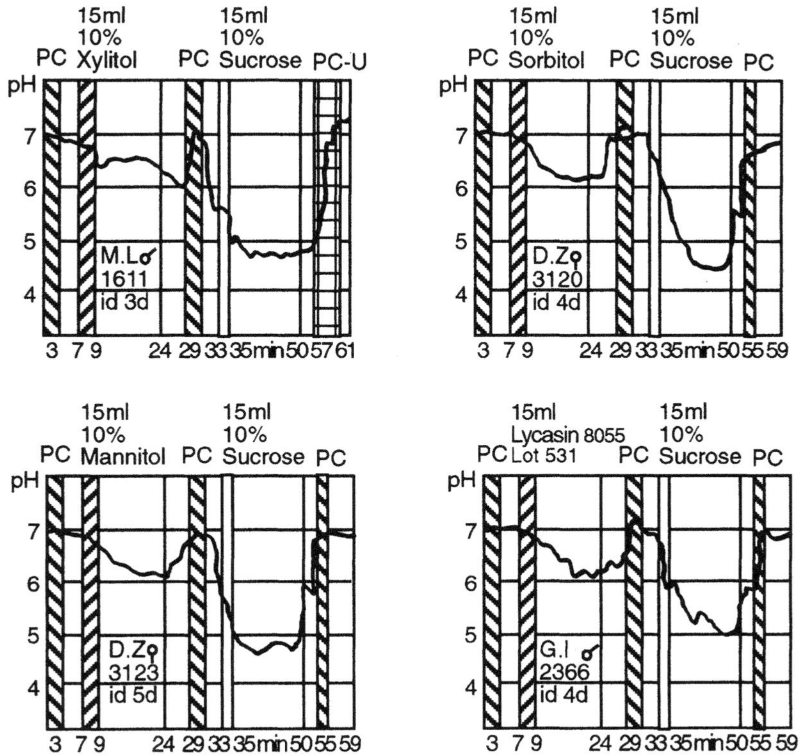

Figure 6.12 Telemetically recorded pH of interdental plaque in different subjects during and 15 min after 2 min rinses with 15 ml each of 10% solutions of xylitol, sorbitol, mannitol and lycasin. 10% sucrose solutions were administered as positive controls. PC = 3 min paraffin chewing; d = days, age of plaque.

Figure 6.13 Tooth friendly logo.

6.6.3 Polyols – safety and tolerance aspect

Polyols are completely safe for use in foods. The acceptable daily intake (ADI) for all polyols is established as 'not specified' by the JECFA (Joint FAO/WHO Expert Committee on Food Additives). This signifies that on the basis of the available data (chemical, biochemical and toxicological), the total daily intake of the substance arising from its use at the levels necessary to achieve the desired effect and from an acceptable background in food, does not, in the opinion of the committee, represent a hazard to health.

Polyols may induce mild flatulence or laxation. This effect is not limited to polyols but is characteristic of many other natural ingredients and foodstuffs, such as fruits, vegetables and other sources of dietary fibre. Different individuals have differing degrees of tolerance. Regular consumption can lead to adaptation and increased tolerance.

The report of the EEC Scientific Committee for Food (SCF) on sweeteners (Opinion of 14 September 1984, updated in December 1987) noted that "consumption of the order of 20 g of polyols per day is unlikely to cause any undesirable laxative symptoms". The SCF further noted that this figure should not be interpreted as a legislative restriction. In the USA a warning about a possible laxative effect is needed only if over 50 g per day are likely to be consumed.

6.7 Applications of polyols

6.7.1 Food industry

In the food industry polyols are used as bulk sweeteners as well as for other technological functions.

6.7.1.1 Bulk sweetener function. Two types of sweeteners are traditionally used in confectionery, sucrose as a crystalline sweetener and glucose syrup which either controls or prevents sucrose crystallization. For sugarless confectionery crystalline bulk sweeteners (maltitol, mannitol, sorbitol and xylitol) and anticrystallizing syrups (sorbitol, maltitol) are widely available. Various types of confectionery can be classified into at least six families of products:

- cocoa based confectionery,
- vitreous hard products: hard candies,
- non-crystallized soft products: jellies,
- partly crystallized soft products: chewing gum, fondant, chews, fudges and so on,

- totally crystallized products: tablets, hard coatings,
- flour confectionery.

6.7.1.1.1 Cocoa-based confectionery. Sucrose is the ideal sweetener for the production of chocolate. The polyol to be used as a substitute should, therefore, be the one presenting the closest technical properties to those of sucrose. Table 6.4 shows the main characteristics of various polyols in 'sugarless chocolate'.

In conclusion it can be said that crystalline maltitol provides the following advantages for chocolate manufacture:

- high sweetness,
- anhydrous crystalline form,
- low hygroscopicity,
- high melting point.

These properties allow pure crystalline maltitol to replace sucrose in high quality chocolate using traditional manufacturing processes. Conventional refining and conching conditions may also be employed.

6.7.1.1.2 Vitreous hard products: hard candies. Various polyols may be used, each with advantages and disadvantages. The optimum choice depends on the process used and type of packaging employed. The use of hydrogenated monosaccharides; e.g. sorbitol and xylitol, requires the use of a depositing line, whereas it is possible to use a forming 'plastic line' with hydrogenated disaccharides or oligosaccharides.

Once the process is finalized, the choice of the optimum polyol is then a function of the type of packaging to be used. In the case of a high moisture barrier packaging, the simplest and cheapest solution consists of cooking a maltitol syrup such as Lycasin® to a temperature which would give a very low (<1%) final moisture content and to wrap the candies just after forming or moulding.

When the packaging does not totally prevent the risk of any water

Table 6.4 Sugarless chocolate

Polyols	Commercial form	Sweetness	Hygroscopicity	Recommended conching temperature (°C)
Sorbitol	Anhydrous	0.6	+	45
Xylitol	Anhydrous	1.0	++	40
Isomalt	Hydrate	0.4	−	45
Lactitol	Hydrate	0.4	−	55
Mannitol	Anhydrous	0.4	−	Same as sucrose
Maltitol	Anhydrous	0.9	−	Same as sucrose

uptake it is necessary to use, together with the maltitol syrup, a crystallizable product which can crystallize on the surface of the candies thus preventing a viscous and tacky layer developing. Blends of Lycasin®, with mannitol, isomalt or lactitol can be used.

6.7.1.1.3 Non-crystallized soft product: jellies. To obtain clear jellies inhibition of crystallization is an absolute must. For this reason the use of non-crystallizable maltitol syrup (such as Lycasin®) is the optimum solution. The manufacure of pectin or gum arabic jellies does not require special attention unlike the manufacture of gelatin/polyol-based jellies which have specialized requirements. In order to achieve a good set of the gelatin, it is necessary:

- to use high bloom gelatins (200 min) and to increase by 20 to 25% the quantity of gelatin used.
- to reach a pH as far from the isoelectric point of gelatin as possible.

6.7.1.1.4 Partly crystallized products. In this type of confectionery, a crystallizing polyol and an anticrystallizing agent can be used to control crystallization. It should be noted that the anticrystallizing agent will have an influence on the control of the water activity.

Combinations of isomalt, lactitol, mannitol and maltitol syrup such as Lycasin® 80/55 may be used in recipes of slightly grained chewies. Lycasin® 80/55 will also act as a plasticizer and control the Aw, for example, of sugarless chewing gums. The manufacture of this type of confectionery does not present major problems provided these different polyols are used in the correct combination.

6.7.1.1.5 Totally crystallized products – respective coating properties of maltitol. The viscosity at saturation for polyols used for hard coating is a quality determining factor. Indeed if the viscosity at saturation is too high, crystallization develops too slowly and cores will be poorly coated. In practice, an upper limit of 200 cps should not be exceeded.

Figure 6.14 shows the viscosity at saturation as a function of temperature. It can be concluded that the coating temperature must be 25°C min for sucrose, 40°C max for sorbitol and 55°C max for xylitol. In theory any temperature could be used for maltitol coating. From a practical point of view 25 to 30°C was found to be optimal (Ribadeau-Dumas, 1994).

6.7.1.1.6 Flour confectionery. In order to produce sponge cake it is possible to use crystalline maltitol and replace the sucrose totally. The product obtained after cooking is similar to that obtained by a traditional sucrose recipe with respect to the viscosity and density of the dough and the colour and volume of finished product. The crumb will also show the correct textural characteristics.

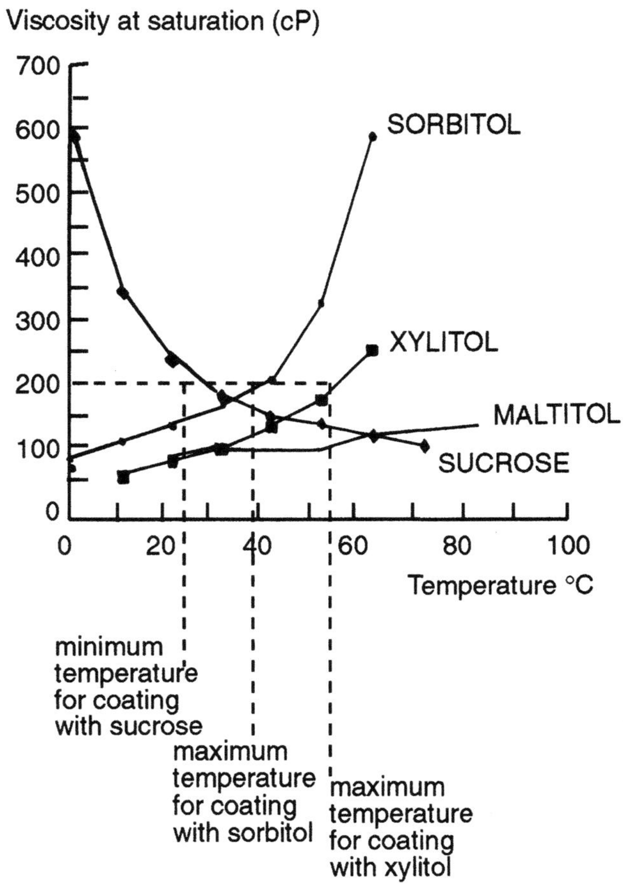

Figure 6.14 Viscosity of saturated solutions.

6.7.1.2 Polyols, food technological function. Owing to price and technological functionality, sorbitol is the most used polyol for reasons other than sweetness.

- As a humectant and stabiliser.
  Foodstuffs contain varying amounts of moisture. If the moisture content is high, they are prone to lose moisture and dry out. Under normal conditions, sorbitol is a very good humectant and can be used successfully in the following applications:
  - Confectionery. In soft confectionery, sorbitol may be used as a humectant at a level of 3 to 15% depending on local legislative requirements. The use of 10% sorbitol in a fondant reduces the water activity by 7%. In sugar-based chewing gum and bubble gum sorbitol

can be used as a humectant and also as a softening agent for the gum base.

- Biscuits, cakes and pastries. Sorbitol may be used at concentrations of 3 to 15% and this will prevent products such as fruit cakes, macaroons and madeleines from drying out. Sorbitol inclusion not only improves the taste, but will also aid in the appearance of the product. Its use in biscuits and cakes containing hazelnuts, coconut and almonds delays rancidity and will thus increase a product's shelf life. It is also possible to replace all the sugar present in a cake or biscuit with sorbitol to produce a product suitable for diabetics.
- Meat-based products, sauces, mayonnaise, fats. Sorbitol is used to stabilize the moisture content and also to improve storage by reducing rancidity.

- As a cryoprotective agent.
  - Fish products. Deep-freezing techniques have become widespread in the fishing industry owing to the extension of fishing zones, the diversity of products to be marketed, rationalization and ease of working on board ship. The use of sorbitol in fish products will optimize the preservation process.

    When used at a level of 3 to 8%, sorbitol will prevent the low temperature denaturation of the protein fibres of the fish. Additionally, sorbitol has a lower sweetness than sugar and will not aid browning.

    Sorbitol is traditionally used in the manufacture of surimi, obtained from the washed flesh of fish, and is also used as a base for numerous preparations including fish 'sausage', crabmeat analogue and kamaboko.
- As an anti-freeze.
  - Ices, ice cream and sorbets. Sorbitol lowers the freezing point of ices and makes them softer and easier to scoop at a given temperature. It will also inhibit the crystallization of other sugars present in the product.

6.7.2 Polyols in pharmaceuticals

Traditionally, sucrose, lactose and starch have been used as excipients in pharmaceutical products. The increased attention paid to oral hygiene has prompted interest in the possibility of replacing cariogenic carbohydrates by excipients that could bring the same technological advantages whilst being non-cariogenic. Since the non-cariogenicity of commercially available polyols is now a well established fact, they are increasingly used in various pharmaceutical forms, mainly for those drugs whose regular consumption could create a dental problem. Presently, in the UK, products like sorbitol, mannitol, xylitol and Lycasin® 80/55 are used in 'sugar free' products. These polyols are in compliance with USP XXII and BP 93.

Liquid non-crystallizable sorbitol (70% DS) is used as a filler and sweetener in cough syrups, decongestant and antihistamine syrups, multivitamin syrups and mouthwashes. Crystalline sorbitol is used as a filler, binder and sweetener in direct compression tablets, e.g. suckable tablets, where its hardness and cooling effect are a major advantage. It is also used as a filler in powder sachets, giving good flowability and rapid dissolution time. Pyrogen-free sorbitol is used in large volume infusions.

Solid mannitol is available in two forms. As a powder, it is used in the manufacture of tablets or capsules (antacids, multivitamin complexes). In granular form, it is most commonly used in direct compression. Pyrogen-free mannitol also has an application in large volume infusions.

Due to its dental properties, xylitol is widely used as a filler and sweetener in gums, hard-boiled candies, in cough syrups and mouthwashes. Lycasin® 80/55 is non-crystallizable, non-cariogenic and has a good chemical–microbiological stability. Its main pharmaceutical applications are in cough syrups, paracetamol suspensions and vitamin syrups. It can also be used for the manufacture of medicated confectionery.

Tables 6.5 and 6.6 summarize the specific properties of polyols which justify their use in food and pharmaceutical applications.

6.8 Conclusion

Polyols are a family of products widely used throughout the world. They can be found naturally in various fruits and vegetables. The polyol range

Figure 6.5 Food applications of polyols

Polyols	Specific properties	Main applications
Crystalline sorbitol	Cooling effect, compressibility, cryoprotective	Chewing gum, tablets, surimi
Sorbitol syrup	Microcrystallization in moulds, surface crystallization, humectancy	Deposited hard boiled candies, dragees, biscuits, cakes, pastry
Crystalline mannitol	Non-hygroscopic	Dusting powder, chewing gum, chewy sweets
Maltitol syrups	Anticrystallizing, plasticity of massecuites, plasticiser, functional substitute for glucose syrups	Hard-boiled and chewy sweets, jellies, chewing gum
Crystalline maltitol	High sweetness, non-hygroscopic, high melting point, crystallization	Chocolate, coated chewing gum
Crystalline xylitol	High sweetness, cooling effect, crystallization	Jellies, chewing gum, coated chewing gum

Table 6.6 Pharmaceutical applications of polyols

Polyols	Specific properties	Main applications
Crystalline sorbitol	Cooling effect, compressibility, rapid dissolving, pyrogen-free grade	Direct compression for suckable and chewable tablets; diluent for sachets; rehydration; vehicle for therapeutics
Sorbitol syrup	Humectant, bulking agent, pyrogen-free grade	Toothpastes, mouthwashes, creams; syrups and liquid medicines; pastilles and lozenges; vehicle for therapeutics; rehydration
Granular mannitol	Compressibility, non-hygroscopic	Direct compression for suckable, chewable and effervescent tablets; diluent for sachets
Crystalline mannitol	Good flowability, high physicochemical stability, pyrogen free grade	Lyophilization carrier; intravenous diuretic
Maltitol syrup	Anticrystallizing agent, chemical stability, high sweetening power	Syrups and all oral liquid forms; pastilles and lozenges
Crystalline maltitol	High sweetening power, high physicochemical stability, molecular weight close to sucrose, good solubility	Diluent for tablets and sachets, coating agent
Crystalline xylitol	High sweetening power, cooling effect, good solubility, physicochemical stability, pyrogen-free grade	Diluent for tablets and sachets; rehydration, vehicle for therapeutics

produced from starch includes sorbitol in liquid or crystalline form, maltitol, maltitol syrup (Lycasin®), crystalline maltitol, mannitol and xylitol. Industrially, polyols are produced by the hydrogenation of selected sugars where the reducing aldehyde or ketone group is converted into a non-reducing alcohol group. They are therefore often referred to as sugar alcohols or polyols.

Polyols are bulk sweeteners which exhibit a pleasant taste and can therefore replace sucrose in many applications. They have good chemical and bacteriological stability. Additionally, they are not fermented by oral bacteria and are therefore not acidogenic. Polyols are 'safe for the teeth'. They have a lower calorific value than sugar and they can act as humectants, stabilizers or cryoprotecting agents by controlling the water activity in the manufacture of stable and fresh food products.

References

Bar, A. (1986) Xylitol. In *Alternative Sweeteners* (eds L.O. Nabors and R.C. Gelardi), Marcel Dekker, New York.

Beaugerie, L., Flourie, B., Franchisseur, C., Pellier, P., Dupas, H. and Rambaud, J.C. (1989) Absorption intestinale et tolérance clinique au sorbitol, maltitol, lactitol et isomalt. *Gastroenterol. Clin. Biol.* **13**, 102 (abstr).
Boussingault, J. (1872) *Ann. Chim. Phys.* **26**, 376.
Caliari, R. (1983) *Manufact. Confection.* **63**, 25.
Den Uyl, C.H. (1985) Lactitol, a new reduced calorie sweetener. In *International Symposium on Polyols and Polydextrose*, Paris, 1985.
Den Uyl, C.H. (1987). Technical and commercial aspects of the use of lactitol in foods as a reduced calorie bulk sweetener. In *Developments in Sweeteners* (ed. T.H. Grenby), Elsevier Applied Science, London, Vol. 3, pp. 65–81.
Devos, F. and Gouy, P.A. (1984) *Patent* EP. 0189704.
Heaton, K.C.F., Robinson, F.D. and Lewin, M. (1980) *Joint Meeting Midland Branch IFST* and *British Soc. Flavorist*, Sutton, Bonnington, p. 157.
Hogg, S.D. and Rugg-Gunn, A.J. (1991) Can the oral flora adapt to sorbitol? *J. Dentistry*, 1991, 263.
Imfeld, T.N. (1977) *Hebd. Odontol. Acta* **21**, 1.
Imfeld, T.N. (1983) Identification of low caries risk dietary components. *Monographs in Oral Science*, Vol 11, Basel, New York, Karger.
Kearsley, M.W., Birch, G.G. and Lian-Loh, R.H.P. (1982) The metabolic fate of hydrogenated glucose syrups. *Starch* **34**, 279.
Le Bot, Y. (1983) Lycasin® for confections. *Manufact. Confection* **63**, 69.
Le Bot, Y. (1989) Polyols for sugarless confectionery: sorbitol, mannitol, xylitol, Lycasin®. In *FIE* **89**, 10, 271.
Le Bot, Y. (1992) Les produits d'hydrogénation de sucres issus de l'amidon, le sucre, les sucres, les édulcorants et les glucides de charge dans les I.A.A. **15**, 362.
Le Bot, Y. (1993) Update on bulk sweeteners polyols, *FIE* **93**, 3, 362.
Quinquenet, S. (1987) 'Contribution à l'étude du polymorphisme du Sorbitol', *Thèse*, UTC Compiègne, France.
Ribadeau-Dumas, G. (1994) *Conf. Seminar ZDS*, Solingen, Germany.
Rugg-Gunn, A.J. (1993) *Nutrition and Dental Health*, Oxford University Press, Oxford, p. 260.
Serpelloni, M. (1985) The food applications of sorbitol, mannitol and hydrogenated glucose syrups. In *International Symposium of Polyols and Polydextrose*, Paris.
Serpelloni, M. (1985) *Patent* EP. 179703.
Serpelloni, M. (1988a) Sugarless confectionery using sorbitol, mannitol and Lycasin® II. *Confection. Prod.* **54**, 332.
Serpelloni, M. (1988b) Sugarless confectionery using sorbitol, mannitol and Lycasin® II. *Confection. Prod.* **54**, 418.
Sicard, P.J. (1982) *Nutritive Sweeteners*, Applied Science, London and New York, p. 145.
Sicard, P.J. and Leroy, P. (1983) *Developments in Sweeteners*, Applied Science, London and New York, Vol. 2, 1.
Sicard, P.J. and Le Bot, Y. (1988) From Lycasin® to crystalline maltitol: a new series of versatile sweeteners. Presented at the *International Conference on Sweeteners–Carbohydrate and Low Calories*, Los Angeles, California.
Strater, P.J. (1988) Palatinit®-an-energy-reduced bulk sweetener derived from saccharose. In *Low Calorie Products* (eds. G.G. Birch and M.G. Lindley), Elsevier Applied Science, London, pp. 63–82.
Taylor, R.L. (1937) *Chem. Metal. Eng.* **44**, 588.
Washuttl, J., Riederer, P. and Bancher, E. (1973) *J. Food Sci.* **38**, 1262.
Weber, W. (1988) Süssungmittel Lycasin® 80/55. Das sollten Si davon wissen. Zucker Süsswren. *Wirtsch.* **41**, 134.
Whitmore, D.A. (1985) Developments in the properties and applications of Lycasin® and sorbitol. *Food Chem.* **16**, 209.
Zeisenitz, S.C. and Siebert, G. (1987) *The Metabolism and Utilisation of Polyols and other Grenby*, Elsevier Applied Science, London, p. 109.

7 Physiology, metabolism and tolerance of digestible and low-digestible carbohydrates

D.M. STOREY and A. ZUMBE

7.1 Introduction

Sugars are soluble carbohydrates of fundamental importance in providing energy for the maintenance of life (Department of Health, 1991). Sucrose is the sugar most widely used as a food in the United Kingdom (Department of Health, 1989) and in the last 150 years its consumption has changed such that it is no longer a minor luxury item but a major dietary constituent. Approximately 14% of the energy requirements of an average British adult are supplied by sucrose (MAFF, 1990), approximately 25% of this being obtained from confectionery products. Although there is no evidence that sugars naturally incorporated in the cellular structure of foods (intrinsic sugars) have adverse effects on health, the recent Department of Health Report on Health and Social Subjects reiterated that, "High intake of non-milk extrinsic sugars can have undesirable effects on dental health and, in the obese, can predispose to undesirable metabolic effects" (Department of Health, 1994). A proposal has thus been made that the population's average intake of extrinsic sugars should not exceed 10% of the total dietary energy (ca. 60 g per day on average; Department of Health 1991). Replacement of the extrinsic sugar sucrose in manufactured foods provides an obvious method of moving towards this goal for health-conscious individuals.

As an example, sucrose is generally considered to be cariogenic and standard milk chocolate usually contains sucrose in the range of 40 to 45% as an added ingredient. However, replacement of sucrose by polyols is possible taking account of a variety of technological criteria such as thermal stability, hygroscopicity, sweetness and water content (Zumbé *et al.*, 1994). Non-caloric, intense sweeteners such as saccharin and aspartame provide an alternative to sucrose when bulking properties are not important, but confectionery manufacture demands such properties thus limiting choice. Polyols (sugar alcohols) are bulk sweeteners which have the advantage of being neither digested nor absorbed to the same extent as sucrose thus being generally accepted as providing less energy per unit mass as well as having reduced cariogenicity (Grenby and Phillips, 1989; Zeisenitz and Siebert, 1987). In healthy humans, partial hydrolysis of disaccharide alcohols occurs in the small intestine but absorption is incomplete; monosaccharide alcohols are also incompletely absorbed. Unabsorbed polyol, or a mixture of unabsorbed hydrolysates together with undigested polyol, contributes to a carbohydrate pool which is available for bacterial fermentation in the colon (Billaux *et al.*, 1991; Livesey, 1992). The ingestion of polyols may thus be associated with increased levels of colonic fermentation and with a dose-related increase in gastrointestinal symptomatology and, if the fermentative capacity of the colon is exceeded, osmotic diarrhoea occurs (Hammer *et al.*, 1989; Rambaud and Flourié, 1994; Saunders and Wiggins, 1981).

This chapter outlines the digestion, absorption, metabolism and tolerance of the digestible carbohydrates starch, sucrose and lactose and gives details of the digestive tolerance of hydrogenated glucose syrup (HGS), e.g. Lycasin®, and of polyols, all of which are partially digestible carbohydrates, especially when they are incorporated into confectionery items such as chewy sweets and chocolate. Particular attention will be given to Lycasin, sorbitol and the disaccharide alcohols lactitol, maltitol and isomalt.

7.2 Carbohydrate digestion

Plant starch is the main dietary carbohydrate, being composed of glucose units held together by 1,4-α-glycosidic linkages with 1,6-α-glycosidic branch points. Starch is hydrolysed by salivary and pancreatic amylases with subsequent hydrolysis of maltose and limit dextrins by a limited number of enzymes such as maltase and α-dextrinase which are intrinsic membrane proteins of the epithelial cells of the brush border of the intestinal mucosa (Figure 7.1).

These enzymes are freely accessible from the lumen and are located in close proximity to active transport sites. Enzyme concentrations are highest in the mid-jejunum and upper ileum where ingested maltose and sucrose are also digested by equivalent processes. Maltase, α-dextrinase and sucrase are components of the sucrase–isomaltase complex and, on contact with these brush border enzymes, substrate saccharides are hydrolysed to release products that are available for absorption (Figure 7.2). Lactase, another brush border enzyme, can hydrolyse several glycosides, the main one being lactose.

Glucose and galactose at low concentration enter the enterocytes rapidly by means of a common sodium-dependent, energy-requiring carrier mechanism. However, when the luminal sugar concentration is high, absorption may be almost entirely passive. Passive absorption of these sugars occurs by solvent drag with water absorption, but in the absence of water movement sugars only move down a concentration gradient and then

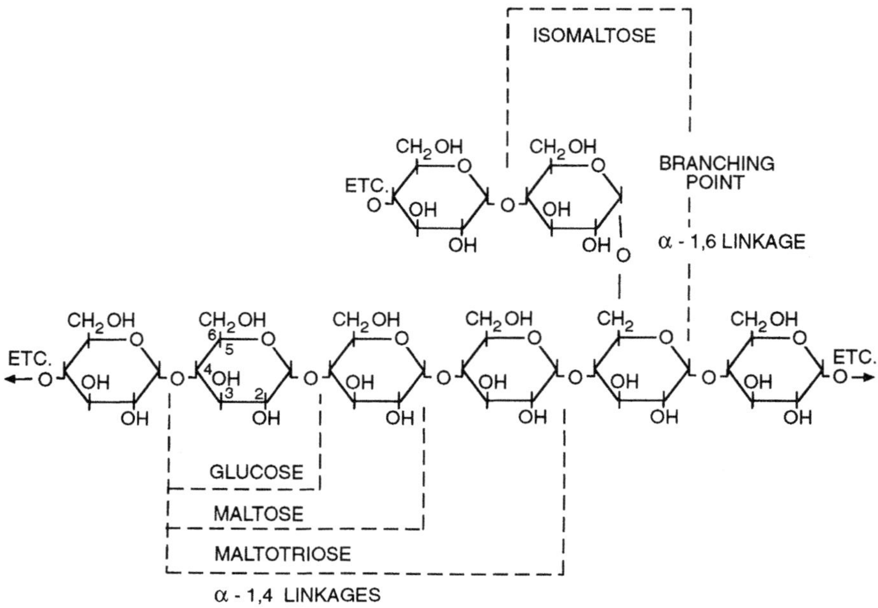

Figure 7.1 Structure and hydrolysis of starch (from Davenport, 1978).

at a rate proportional to the magnitude of the gradient. Some 10% or more of the absorbed glucose may be metabolised within the absorptive cells but the remainder makes its exit from the cytosol into the intracellular space partly by diffusion but mostly via a sodium-independent carrier located at the basolateral membrane (Johnson, 1991).

Fructose is also exceptionally well absorbed but in this case by facilitated diffusion only, because, unlike the glucose–galactose carrier, the fructose carrier cannot be energised for active transport. Some control over the rate of sugar absorption is possible because stomach motility and emptying can be slowed in response to negative feedback signals from the duodenum and proximal jejunum; nevertheless, both glucose and fructose may remain transiently in the luminal contents. Although these molecules are osmotically active, the speed with which they are absorbed normally ensures that water is not required to bring the intestinal contents to isotonicity and thus osmotic diarrhoea does not occur. The capacity of the human small intestine to absorb free sugars is indeed enormous and it has been estimated that hexoses equivalent to 22 pounds of sucrose could be absorbed daily! (Johnson, 1991).

The concentration of disaccharidases can increase in response to certain changes in the diet, sucrase being inducible by an increase in dietary sucrose although lactose, galactose and maltose do not appear to stimulate

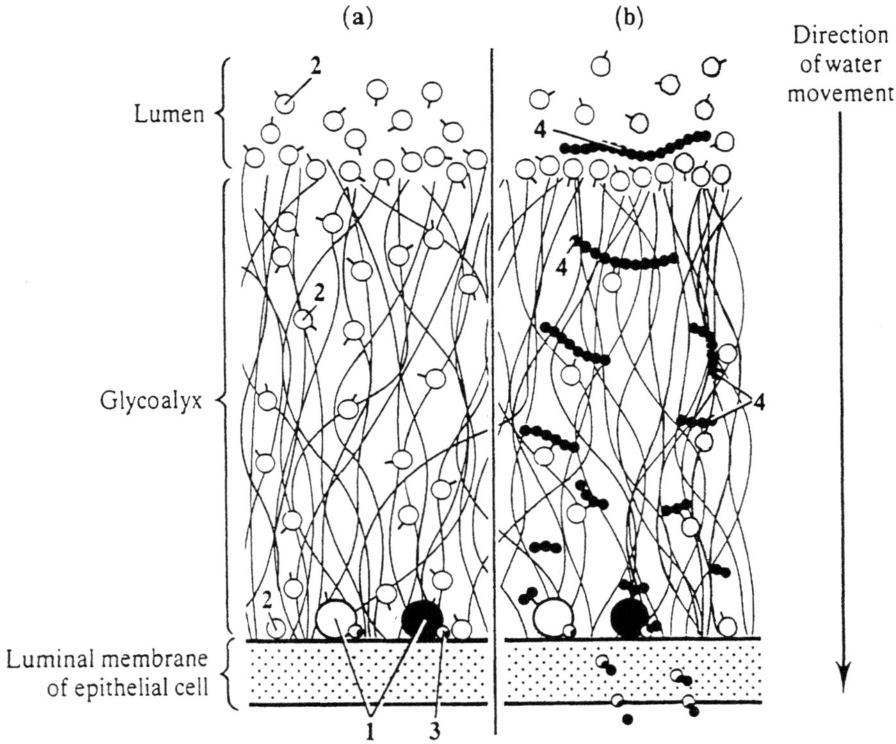

Figure 7.2 Membrane digestion and absorption of sugars (from Ugolev, 1972/74). (a) Distribution of enzymes. (b) Relationship of enzymes, carriers and substrates. 1 = enteric enzymes; 2 = adsorbed enzymes, e.g. pancreatic amylase; 3 = carriers; 4 = substrates.

similar effects. However, disaccharidase deficiencies are known to occur, being associated with well-documented malabsorption syndromes. Lactose intolerance, which may be due to congenital or later acquired deficiency of lactase, is associated with colonic fermentation of unhydrolysed lactose with accompanying flatulence, abdominal cramps and the passage of lactose, lactic acid and fatty acids in the stools. Diarrhoea occurs because of the high osmotic pressure of the colonic contents and the stimulation of peristalsis by acid products of fermentation. Rarely, deficiencies of sucrase and α-dextrinase may occur in children but maltose intolerance appears not to occur because of the existence of several enzymes capable of hydrolysing maltose (Figure 7.3). Intolerance of glucose and galactose does occur in rare instances, apparently due to the absence of the glucose carrier and patients may be helped to recover by feeding them only fructose, whereupon symptoms disappear.

The ingestion of poorly absorbed carbyhydrate leads to the development of a symptom complex of which diarrhoea is the main feature (Menzies,

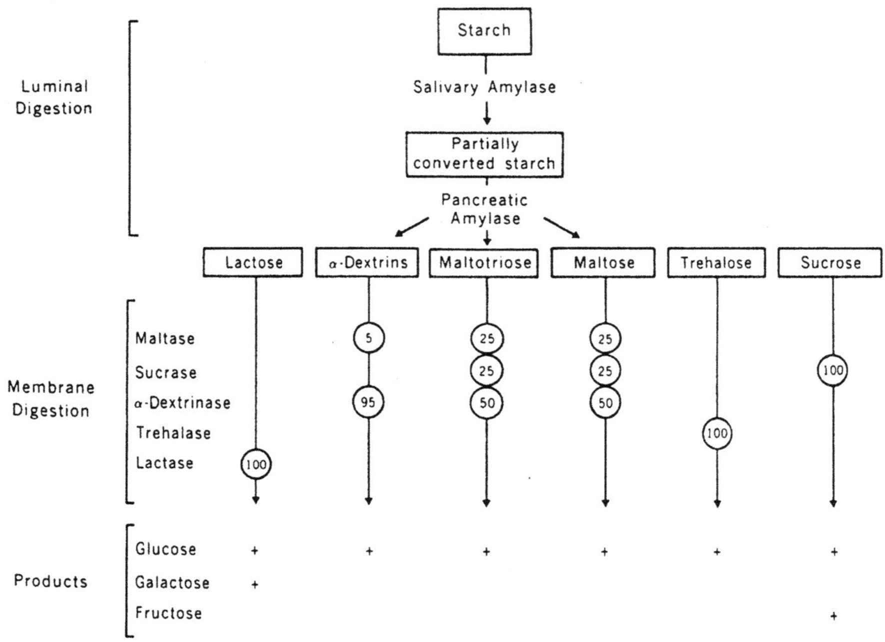

Figure 7.3 Summary of the luminal digestion of starch and membrane digestion of disaccharides and oligosaccharides (from Johnson, 1991).

1983). Diarrhoea may be defined as a stool volume greater than 500 cm^3/day and is usually accompanied by increased frequency of stool passage as well as by an increased fluidity of the motions (Figure 7.4). Although the colon has a tremendous ability to absorb water, diarrhoea may occur if it is overwhelmed by the volume of fluid delivered from the small intestine or if the ileum delivers unabsorbed sugars which are osmotically active, thus reducing the absorptive capacity of the colon. In fact very few naturally occurring foods contain a high enough concentration of unabsorbable sugar to cause diarrhoea although other symptoms such as bloating, colic and flatulence may occur.

7.3 Low digestible carbohydrates

Polyols are quite closely related to sugars, with the reducing group of the sugar being replaced by a hydroxyl group. Hydrogenated monosaccharides include sorbitol, mannitol, xylitol and erythritol, and hydrogenated disaccharides include maltitol, lactitol and isomalt, whereas Lycasin® is a mixture of hydrogenated saccharides and polysaccharides (Figure 7.5). In the EC, all polyols and HGS are considered as food additives (Le Bot,

Consequences of Sugar Malabsorption = FERMENTATIVE DIARRHOEA

MECHANISM | **CLINICAL FEATURES**

Due to Sugar Retention in Gut

SMALL INTESTINE

Osmotically Induced Fluid Entry → Intestinal Hurry / Griping pains — Increased Peristalsis

LARGE INTESTINE

Bacterial Sugar Fermentation

- Enhanced Osmotic Effect → Watery Diarrhoea → Fluid and Electrolyte loses
- Generation of
  - Organic acids (lactic, etc)
  - Gases (CO_2, H_2, CH_4 etc) → Acid stools (pH < 6) / Flatulence / Abdominal Distension

Figure 7.4 Consequences of sugar malabsorption (from Menzies, 1983).

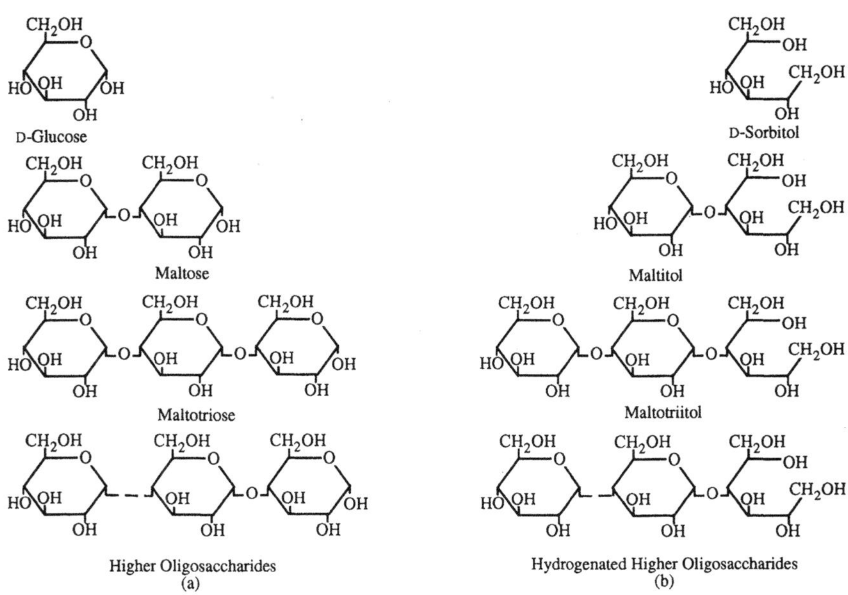

Figure 7.5 Chemical structure of polyol precursors and polyol components of hydrogenated glucose syrup (from Moscowitz, 1980). (a) Higher oligosaccharides, (b) hydrogenated higher oligosaccharides.

1993). Lycasin® (HGS), maltitol, mannitol and sorbitol are manufactured directly or indirectly from starch, the other polyols being manufactured from other raw materials such as milk, beet and cane sugar and corn cobs (Billaux *et al.*, 1991). Further details of these processes are to be found elsewhere in this volume.

7.3.1. *Polyol content of natural foods*

Polyols (mainly sorbitol, mannitol and xylitol) occur naturally in a large number of fruits and vegetables. In fresh fruits up to approximately 8% polyols can be found, but in certain dried fruits the polyol content increases up to 21% (Table 7.1). The polyol content of vegetables is lower than that of fruits but concentrations up to 0.8% have been recorded in some fresh vegetables with up to 1.5% in certain mushrooms (Table 7.2). Based on data published by the Food and Agriculture Organisation (FAO), Vincent (1989) estimated the theoretical total annual consumption of polyols in fruits as 1.162×10^6 tonnes of sorbitol, 4.8×10^5 tonnes of xylitol and 2.4×10^3 tonnes mannitol. He also estimated the intake of polyols from added sweeteners in food and pharmaceutical products to be 2.75×10^5 tonnes, meaning that the intake of polyols added to food is less than 10% of the total polyol intake.

Table 7.1 Polyol content of various fruits (BLS, 1990)

Fruit	Polyol content (mg 100 g^{-1})	Polyol content (%)
Rowanberry, dried	20 984	21.0
fresh	8534	8.5
jam	4080	4.1
Plums, dried	14 872	14.9
fresh	2975	3.0
jam	5366	5.4
juice	1511	1.5
Pear, dried	8417	8.4
fresh	1624	1.6
juice	877	0.9
Apricot, dried	4296	4.3
fresh	792	0.8
juice	437	0.4
Apple, dried	3202	3.2
fresh	585	0.6
Peach, dried	1787	1.8
fresh	267	0.7
juice	182	0.2
Orange, dried	1070	1.1
fresh	190	0.2
juice	109	0.1
Wild fruits, fresh	1261	1.3

Table 7.2 Polyol content of various vegetables (after Souci *et al.*, 1989)

Vegetable	Polyol content (mg 100 g^{-1} wet weight)	Polyol content (%)
Ringed boletus	1440	1.4
Mushrooms	1150	1.2
Honey mushrooms	1090	1.1
Parsley (root), fresh	820	0.8
Sweetcorn, fresh	200	0.2
Carrot, fresh	187	0.2
Leek, fresh	150	0.2
Celery, fresh	110	0.1
Cucumber, fresh	70	0.07
Cauliflower, fresh	26	0.03
Pumpkin, fresh	26	0.03

7.3.2 Polyol content of no-sugar-added or energy-reduced products in the marketplace

Polyols and HGS are used in the food, cosmetic and pharmaceutical industries, their main usage being as bulk sweeteners (giving both volume and sweetness) in the production of sugar-free, calorie-reduced or tooth-friendly foods and foods for diabetics. The technological handling properties of many polyols used in the manufacture of foods are comparable with those of sugar, enabling increased use of polyols as alternative bulk sweeteners to take advantage of their hypocaloric and hypocariogenic properties. They are also used in small amounts as humectants. The most important food category for the use of polyols is confectionery (chewing gum, candies, compressed tablets and chocolate). Table 7.3 gives a review of the typical polyol content of various sugar-free/no-added-sugar and some energy reduced products.

7.3.3 Digestion and absorption of hydrogenated glucose syrups and polyols

7.3.3.1 Hydrogenated glucose syrups. These are first hydrolysed by α-amylase producing di- and oligosaccharides, after which intestinal disaccharidases split dimers containing only glucose to free glucose which is easily absorbed and utilised by the body. However disaccharides containing sorbitol linked to glucose are hydrolysed slowly with little resultant absorption (Dahlqvist and Telenius, 1965). Lycasin 80/55® is a hydrogenated glucose syrup with a particular, defined composition (Sicard and Leroy, 1983) and more than 96% of it is broken down to glucose and sorbitol, the remainder being maltitol (Verwaerde and Dupas, 1982). Most of the glucose is absorbed but some of the sorbitol is not and is presented

Table 7.3 Typical polyol content of sugar-free/no-added-sugar products and some energy reduced products (Willibald-Ettle, 1994)

Product	Typical polyol content (%)
Candies	
– hard	98
– soft	70–75
Compressed tablets	78–99
Chewing gum	up to 70
– coated chewing gum	up to 80
Chocolate	
– bar	39–44
– light recipe	30
Baked goods	
– biscuits	20–30
– sponge cake	20–30
Ice cream	15
– light	10
– light (other recipe)	4

intact to the large intestine where it is fermented. Maltitol may be further hydrolysed by α-glycosidase and isomaltase of the brush border but again, some remains intact and enters the colon where fermentation occurs. Very little faecal excretion of polyols occurs after the ingestion of Lycasin®. The sorbitol content of Lycasin® is, therefore, a major determinant of its properties and the digestion of sorbitol and the other polyols are dealt with further below.

7.3.3.2 Sorbitol. This is absorbed across the small intestine mucosa passively by facilitated diffusion (Lauwers *et al.*, 1985), after which it is oxidised slowly to D-fructose by sorbitol dehydrogenase in the liver then entering the fructose-1-phosphate pathway without stimulating a glycaemic response (see below). However, sorbitol has a relative molecular mass of 182 D which is near the limit for passive diffusion (Dwivedi, 1978) and it is clear that absorption is incomplete (especially at low concentration) and that residual sorbitol enters the large bowel where it is fermented, this becoming apparent by the appearance of hydrogen in exhaled breath. For example, a significant increase in the amount of hydrogen excreted in breath could be detected when as little as 5 g sorbitol was eaten (Hyams, 1983). Interestingly, Krüger *et al.* (1992) showed that in rats the digestion of sorbitol lacked dose dependence although the physiological response, in the form of intestinal water movement strongly correlated with the dose administered.

Beaugerie *et al.* (1990) demonstrated that 79% of a 30 g dose of sorbitol remained unabsorbed when ingested as an oral solution, although the degree of malabsorption was affected by several factors including, importantly, the composition of the solution in which the dose was administered. The mean percentage of sorbitol absorbed in the small intestine was significantly higher in pure sorbitol doses than in those containing maltitol and Lycasin® 80/55. (In contrast, the mean percentage of total maltitol metabolised was not significantly different for pure maltitol and maltitol contained in Lycasin® 80/55.) The vehicle of sorbitol ingestion was also considered by Zumbé and Brinkworth (1992) who studied the acute consumption of 20 g sorbitol incorporated into milk chocolate. They demonstrated incomplete sorbitol absorption that was manifest as osmotic diarrhoea; however, the subjects in their study showed increased tolerance if the dose was spread out over the day. Studies in which [U-^{14}C]sorbitol was given either to normal rats (Ertel *et al.*, 1983) or germ-free rats (Wursch *et al.*, 1990) demonstrated 55% and 51% residual activity respectively in the gut contents, most being in the caecum. Beaugerie *et al.* (1990) were able to further show that the faecal excretion of sorbitol is negligible indicating that it is almost completely digested by the colonic flora.

7.3.3.3 Mannitol. Mannitol, like its isomer sorbitol, is only partially and slowly absorbed in the small intestine. Unabsorbed mannitol passes into the colon where it is fermented by bacteria. Nasrallah and Iber (1969) reported that on average 35% of an ingested dose of [U-^{14}C]mannitol is excreted in the stools and 20% in the urine. The fact that very little $^{14}CO_2$ was expired by subjects in the study referred to above suggests that very little mannitol is metabolised in the tissues and what is absorbed is excreted in the urine.

7.3.3.4 Xylitol. This differs from sorbitol and mannitol in that it is a pentitol. Absorption from the small intestine occurs by diffusion after which it is mostly metabolised in the liver to D-xylulose, which is then phosphorylated and converted to hexose-6-phosphate and eventually glucose (Wang and Van Eys, 1981).

7.3.3.5 Erythritol. This differs from all the other polyols so far considered in that it is a four-carbon sugar alcohol. It occurs naturally (Shindou, 1989) but can also be produced in large amounts by fermentation of glucose and sucrose with *Aureobasidium* sp. (Goossens and Roper, 1993). Using [U-^{13}C]-labelled erythritol Hiele *et al.* (1993) reported that erythritol is rapidly absorbed in the small intestine by passive diffusion but that no $^{13}CO_2$ is excreted in the exhaled breath. Oku and Noda (1990) found that when erythritol was administered to humans at 0.1 g kg^{-1} body weight, 80% was excreted in the urine and 6% in the expired air. Presumably the colonic bacteria of Oku and Nodas' subjects had adapted to metabolise whatever polyol entered the colon, whereas in Heile's subjects they had not? Erythritol has 75% of the sweetness of sucrose so it might be useful as a sweetener which because of its rapid absorption will not stimulate gastrointestinal symptomatology and, because it is mostly excreted in the urine without a post-prandial insulin response, might be suitable for diabetics. However little work is available in the public domain concerning this sugar alcohol and diuretic effects might prove to be a problem.

7.3.3.6 Lactitol. Lactitol (4-O-β-D-galactopyranosyl-D-sorbitol) is a disaccharide alcohol that requires hydrolysis prior to absorption and if completely hydrolysed yields equimolar amounts of galactose and sorbitol. However, Nilsson and Jagerstad (1987) used human intestinal biopsies *in vitro*, to demonstrate that lactitol is a poor substrate for lactase which showed an activity of only 1% of that towards lactose which itself is around 25% of that of sucrose. Furthermore, lactitol is not absorbed intact from the small intestine to any great extent (Metzger *et al.*, 1988; Patil *et al.*, 1987). Beaugerie *et al.* (1991) showed that up to 84% of a 20 g dose of lactitol given in an iso-osmotic solution remained unabsorbed. Using [U-^{14}C]lactitol Grimble *et al.* (1988) showed that 62.9% of a 20 g dose was excreted as $^{14}CO_2$ via the lungs whilst only 6.5% and 2% were recovered from the urine and faeces, respectively. This suggests that lactitol is extensively fermented in the human colon and that a significant proportion of the bacterial metabolites are available for colonic absorption and ultimately metabolised in the tissues.

7.3.3.7 Maltitol. Maltitol (4-O-α-D-glucopyranosyl-D-sorbitol) is hydrolysed by enzymes of the small intestine brush border (Lian-Loh *et al.*, 1982; Rosiers *et al.*, 1985; Wursch and Del Vedovo, 1981; Wursch *et al.*, 1990; Ziesenitz and Siebert, 1987), particularly by the α-glycosidase of the sucrase–isomaltase complex (Zunft *et al.*, 1983). This enzyme has a much lower affinity for maltitol than it does for either sucrose or maltose (Kamoi, 1975), releasing equimolar amounts of glucose and sorbitol. However maltitol is also hydrolysed by isomaltase (Zeisenitz, 1986). In their *in vitro* study, Nilsson and Jagerstad (1987) reported that maltase has 10% activity towards maltitol compared to maltose. Nevertheless, Beaugerie *et al.* (1991) calculated that about 90% of ingested maltitol is digested with subsequent absorption of approximately 70% of the sorbitol released and nearly all the glucose. Hence they suggested that approximately 75% of the ingested dose is digested with the remaining 25% entering the colon as maltitol and sorbitol (2:3). According to Beaugerie *et al.* (1990) any maltitol reaching the small intestine is almost completely digested by the colonic bacteria and almost none appears in the faeces. On the other hand, Rennhard and Bianchine (1976) gave doses of [U-^{14}C]-

labelled maltitol to volunteers and detected 3.6% and 4.6% in the urine and stools, respectively.

Most of the other research concerning the digestion and absorption of maltitol comes from non-human studies but throws some light on what might occur in the human gut. Wursch *et al.* (1990) showed that in rats and mice 40% of ingested maltitol reached the colon, rather more than is proposed for humans. Contrary to this, Oku *et al.* (1971) claimed that although 5% of [U-^{14}C]maltitol ingested by rats was absorbed directly, being rapidly excreted in the urine, the greater part passed into the colon where it was fermented rather than being hydrolysed by digestive enzymes in the small intestine. Similar results were also obtained in rat studies by Lian-Loh *et al.* (1982). Krüger *et al.* (1992) studied the kinetics of gastrointestinal transit of different loads of maltitol in rats and showed that its digestibility and absorption in the small intestine depends inversely on the dose ingested and also on the maltitol cleavage product sorbitol being very slowly absorbed. In general, however, overall support can be given to the conclusion of Nilsson and Jagerstad (1987) that, unlike lactitol and isomalt, significant amounts of maltitol can be digested and utilised by humans.

7.3.3.8 Isomalt. Isomalt (D-glucosyl-α (1-1)-D-mannitol and D-glucosyl-α-(1-6)-D-sorbitol) is cleaved very slowly by the small intestinal hydrolases (Grupp and Siebert, 1978). The α-glycosidase of the sucrase–isomaltase complex hydrolyses isomalt to yield 50% glucose and 25% each of sorbitol and mannitol. Grupp and Siebert concluded that, in humans, the relative rate of isomalt cleavage by α-glycosidase was 2% that of maltose. Nilsson and Jagerstad (1987) used human intestinal biopsies *in vitro* to demonstrate that the hydrolysis of isomalt is about 10% of the expected rate for palatinose but only 1 to 3% of that towards isomaltose. Using mucosal homogenates, purified brush border enzymes and everted segments of rat jejunum, Goda *et al.* (1988) demonstrated that hydrolysis of isomalt was 6 to 7% of that of sucrose. More recently, Kruger *et al.* (1991) demonstrated that in rats isomalt hydrolysis was more complete when administered at lower doses than at higher doses. In pigs, about 60% of ingested isomalt passes the end of the small intestine (Van Weerden and Huisman, 1993a, b) but none appears in the faeces, indicating that it is completely digested by bacterial fermentation in this species. In humans, Grupp and Siebert (1978) found that very little isomalt was excreted in either the faeces or urine, levels being 0.2% and 0.1%, respectively.

7.3.4 Insulin response following ingestion of low digestible carbohydrates

Rising plasma glucose levels following glucose absorption from the alimentary canal stimulate the release of insulin from the β-cells of the islets of Langerhans via the mediator GIP. Sucrose is known to elicit a higher plasma glucose response and thus a higher insulin response than complex carbohydrates (Crapo *et al.*, 1976). The rate of absorption into the blood of glucose from glucose syrups is in the same order as that from glucose *per se*, but compared to sucrose there are some advantages in that glucose syrups fail to stimulate the rise in plasma lipids seen after high sucrose diets and consequently a smaller amount of fat is laid down in the liver (Lian-Loh *et al.*, 1982). It was subsequently determined that the total serum insulin response over the 2 h period following the consumption of Lycasin® and maltitol syrups was significantly smaller than that following the consumption of D-glucose. Hydrogenated glucose syrups thus behaved differently from normal glucose syrups resulting in a smaller insulin response, possibly because of the non-availability of some covalently bound glucose in HGS (Lian-Loh *et al.*, 1982).

Lactose intolerant people are also known to have a flat blood glucose curve after the ingestion of lactose because they lack lactase, the enzyme which is necessary to hydrolyse lactose. For similar reasons polyols are also absorbed minimally from the small intestine and they typically lead to glycaemic and insulin responses which are characteristically lower than those occurring after the consumption of either glucose or sucrose. For this reason HGS and polyols have both been used for several years to sweeten products for diabetics who need to avoid large post-prandial rises in glucose and insulin.

In a prospective double-blind controlled crossover study on ten healthy males aged 21 to 30 years, Thiébaud *et al.* (1984) investigated the acute effects of oral ingestion of 30 g loads of isomalt and sucrose on plasma glucose and insulin. After sucrose, plasma glucose and insulin increased markedly in the first 60 min whereas after isomalt, plasma glucose and insulin levels increased only slightly during the first 90 min. The maximum increment for plasma glucose was 2.78 ± 0.29 versus 0.33 ± 0.29 mmol l^{-1} after isomalt. For plasma insulin, the maximum increment was 22.4 ± 7.2 after glucose versus 3.4 ± 1.7 μU ml^{-1} after isomalt showing a sevenfold to eightfold difference in maximum values for both glucose and insulin levels.

Maltitol also causes a lower rise in blood glucose and insulin levels than a corresponding oral dose of glucose or sucrose. Felber *et al.* (1987) conducted a study in which eight normal subjects consumed 30 g of either maltitol or sucrose on two separate days with an interval of at least a week. Plasma glucose and insulin levels both peaked 30 min after ingestion of the load for both sugars, but the oral load of maltitol produced markedly less stimulation of carbohydrate oxidation than an equivalent load of sucrose. The peak of the glucose concentration was significantly smaller after maltitol than after sucrose (2.1 ± 0.4 versus 3.8 ± 0.4 mmol l^{-1}) as was the peak of the insulin concentration after maltitol and sucrose, these being

9.3 ± 2.7 versus 25.5 ± 5.0 μU ml^{-1} respectively. Recently, Tsuji *et al.* (1990) also showed that after the intake of 10 g of either glucose or maltose by each of five subjects, their serum glucose and insulin levels reached a maximum at 30 min post-consumption. However, there was only a slight glycaemic response after taking maltitol and neither serum glucose nor insulin levels showed any response after taking sorbitol.

As in the previous examples, plasma glucose levels did not increase following the consumption of lactitol (0.5 g kg^{-1}) by six healthy subjects aged 19 to 38 years and eight patients with liver cirrhosis (Metzger *et al.*, 1988). Mean basal plasma glucose levels were 5.8 and 4.4 mmol l^{-1} and the maximum increments following lactitol were +1.4 and +0.6 mmol l^{-1} for cirrhotic and healthy subjects, respectively.

7.4 Intestinal fermentation and tolerance

7.4.1 Intestinal fermentation

In healthy humans, partial hydrolysis of disaccharide polyols occurs in the small intestine but absorption is incomplete and a mixture of hydrolysates and undigested polyol passes to the large bowel and is subject to bacterial fermentation (Billaux *et al.*, 1991; Livesey, 1992). Monosaccharide alcohols are also incompletely absorbed and contribute to a carbohydrate pool which also includes unhydrolysed polyol, undigested starch, dietary fibre and endogenous mucins, and which is available for fermentation. The products of this fermentation are the gases H_2, CO_2, CH_4 (in about one-third of the population) and organic acids, mainly the short chain fatty acids (SCFA) acetate, propionate and butyrate but also including succinate, formate and lactate as well as ethanol. Miller and Wolin (1979) calculated that in humans SCFA account for about 72% of the energy content of luminal carbohydrate excluding that used by the bacteria for growth. SCFA concentrations rise during the passage of intestinal contents from the terminal ileum (ca. 4 mM) into the colon (80 to 90 mM) from where they are absorbed by the epithelial cells at rapid rates ranging from 6 to 12 μmol cm^{-2} h^{-1} (McNeil *et al.*, 1978). However, Hammer *et al.* (1989) demonstrated that the organic acids produced in diarrhoea induced by carbohydrate (lactulose) malabsorption were only partially absorbed in the colon and that unabsorbed organic acids, being anions, obligated the accumulation of inorganic cations ($Na^+ > Ca^{2+} > K^+ > Mg^{2+}$) in the diarrhoeal fluid. In the case of low doses of lactulose, diarrhoea was induced by unabsorbed organic acids and associated cations whereas with larger doses of lactulose, unmetabolised carbohydrate also played a major role. Overall, however, in the case of lactulose the net effect of bacterial metabolism and the effect of partial absorption of organic acids on stool

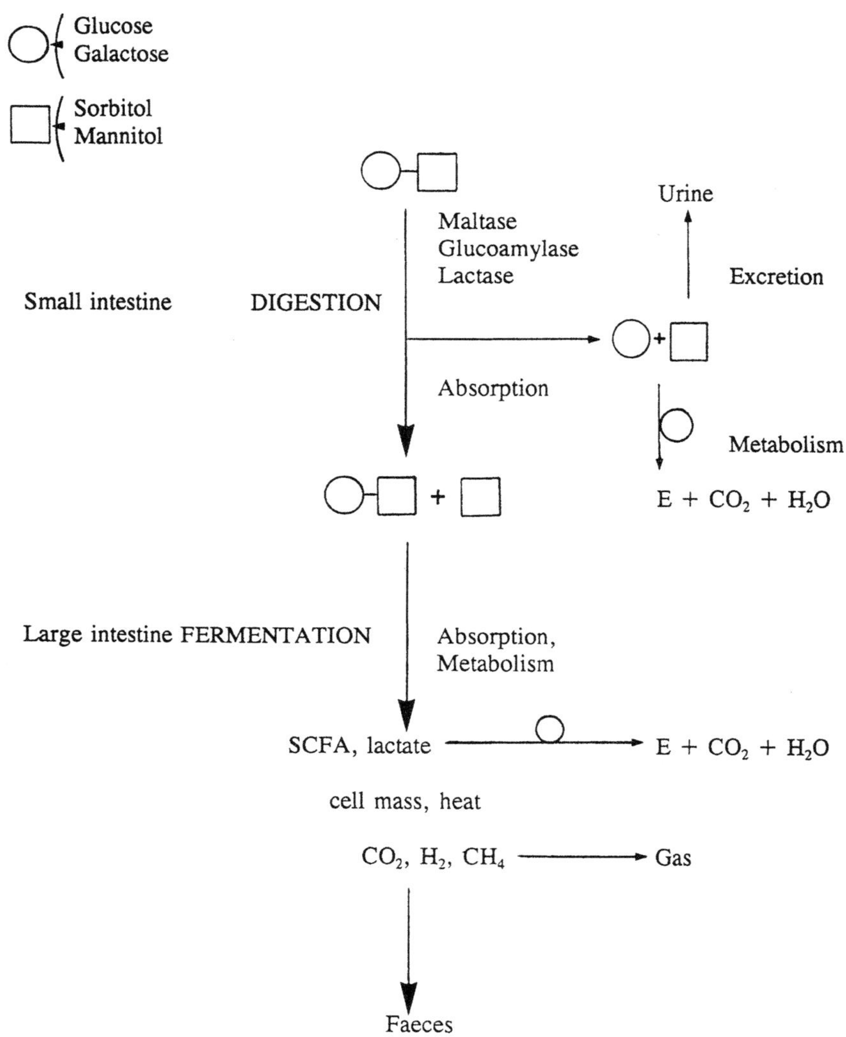

Figure 7.6 Metabolic fate of disaccharide alcohols (from Wursch and Anantharaman, 1989).

water output was dose dependent. The same is almost certainly true following the ingestion of other partially digestible carbohydrates such as the disaccharide alcohols. Those SCFAs which are absorbed then enter the general metabolism and Figure 7.6 (Wursch and Anantharaman, 1989) elegantly summarises the metabolic fate of the disaccharide alcohols.

SCFAs have effects other than providing an energy source and influencing the movement of inorganic ions and water. Some can stimulate either colonic secretion, motility or epithelial cell division. For example an increase in Cl^- secretion is produced by propionate, although not by acetate, perhaps by a mechanism similar to that by which secretagogues act in the small intestine (Johnson, 1991). It has been proposed that SCFAs activate endocrine-like cells which release peptides and/or amines into the interstitial fluid. As a result, nearby neurones are excited and a secretion elicited which releases acetyl choline close to the muscarinic cholinergic receptors on the basolateral membranes of secretory colonic cells (Figure 7.7). SCFAs also stimulate the growth of the colonic mucosa, butyrate being more potent than either acetate or propionate.

The ingestion of polyols, which are poorly absorbed, is thus associated with an increased concentration of carbohydrate entering the large bowel and with increased levels of colonic fermentation. If the fermentative capacity of the colon is exceeded osmotic diarrhoea occurs (Hammer *et al.*, 1989; Rambaud and Flourié, 1994; Saunders and Wiggins, 1981) and is associated with a consequent increase in gastrointestinal symptomatology including colic, flatus and borborygms. Although polyols have a laxative effect when consumed in large amounts, this effect cannot be considered to be toxic and is modified by the colonic fermentation of the responsible sugar.

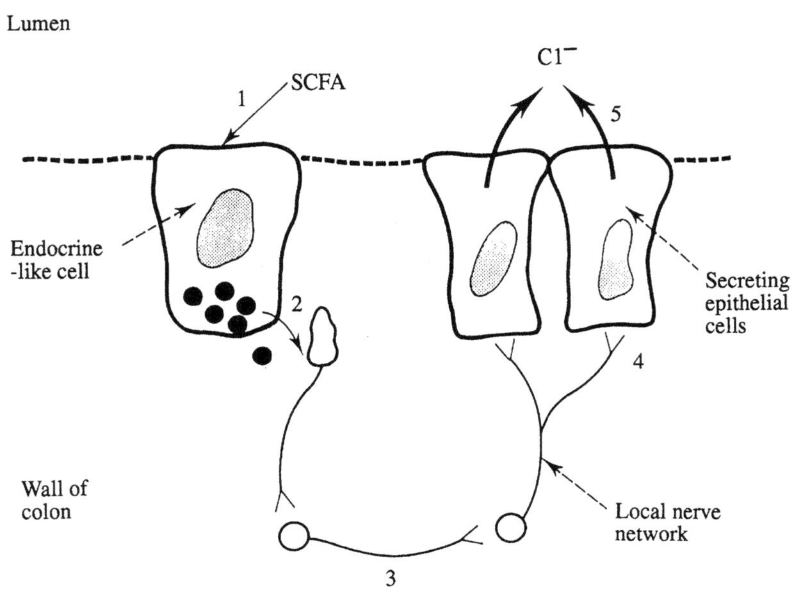

Figure 7.7 Model illustrating some effects of SCFAs on colonic secretion (from Sanford, 1992). (1) SCFAs act on endocrine-like cells releasing peptides and/or amines which (2) activate a local nerve plexus which (3) releases acetyl choline which (4) binds to muscarinic receptors on colonic epithelial cells resulting in (5) increased Cl^- and fluid secretion.

7.4.2 Stool consistency, laxation and diarrhoea

Grimble *et al.* (1988) found that 2/6 adapted subjects had diarrhoea after consumption of lactitol on an empty stomach. However, other studies have shown that if the dose is fractionated or is taken with meals laxativity is less pronounced (Patil *et al.*, 1987; Van Es *et al.*, 1986). Similarly, Beaugerie *et al.* (1989) did not observe laxation either in six healthy subjects after the intake of 20 g maltitol into an empty stomach or after the ingestion of three times higher doses when consumed by adapted subjects after eating three meals. The same can be said of isomalt (Beaugerie *et al.*, 1989; Thiébaud *et al.*, 1984) when consumed under similar conditions. However, there is no doubt that the ingestion of large enough doses of polyols can have a laxative effect. Gryboski (1966) found that sorbitol incorporated into candies caused osmotic diarrhoea when ingested at levels of 20 to 50 g. In a study in which volunteers were fed diets supplemented with either lactitol or saccharose, Van Es *et al.* (1986) claimed that sorbitol promotes osmotic diarrhoea whereas lactitol is fermented in the large intestine resulting in increased gas formation and the production of an increased faeces biomass. From a therapeutic standpoint, Heitland and Mauersberger (1988) found that 20 g of crystalline lactitol powder and 30 cm^3 lactulose syrup were efficacious in increasing stool frequency. Similarly, Goovaerts and Ravelli (1993) reported the efficacy of lactitol for the treatment of chronic functional constipation, but found that with a dose of 20 g lactitol per day up to 46% of subjects experienced adverse gastrointestinal side effects such as abdominal cramps/pain, meteorism, flatulence and diarrhoea.

Thus, ingestion of sufficient amounts of non-absorbable or poorly absorbable, low molecular weight carbohydrates such as polyols may induce osmotic diarrhoea which originates in the small intestine as a result of profuse water and electroyte secretion into the duodenum and proximal jejunum (Rambaud and Flourié, 1994). The increased flow rate exceeds the adaptability of the intestine's small diameter with a consequent shortening of oro-caecal transfer time which further decreases the hydrolysis and absorption of the already poorly absorbed carbohydrate (Krüger *et al.*, 1992; Rambaud and Flourié, 1994). When the non-absorbed carbohydrates enter the colon they are fermented as outlined in section 7.4.1 above. The colonic metabolism of polyols into readily absorbable short chain fatty acids (SCFA) thus reduces the osmotic load which abrogates the potential diarrhoea which would otherwise have resulted from the high ileal output. McNeil (1984) pointed out that "all that enters the large intestine is not lost" and ". . . dietary carbohydrate, rather than leading to diarrhoea, is fermented in the colon and enhances the absorption of fluid and electrolytes".

The overall laxative effect is thus determined by the slow, passive absorption rate of pentitols (xylitol) and hexitols (sorbitol and mannitol) and to the incomplete hydrolyis and absorption of polydextrose and disaccharide alcohols (isomalt, lactitol and maltitol). Sugars and sugar alcohols are osmotically active and thus influence the osmoregulatory capacity of the bowel. Although the colon is endowed with an extraordinary capacity to conserve fluid and electrolytes, its absorptive capacity can be exceeded in certain circumstances. Sugars such as glucose and fructose initially result in water transport into the gastrointestinal tract but they are rapidly absorbed by active and passive mechanisms, and so their osmotic effect is quickly dissipated. Xylitol, sorbitol and mannitol have the strongest osmotic effect of all the polyols followed by polydextrose, lactitol, isomalt and maltitol according to their molecular weight (Rambaud and Flourié, 1994)

7.4.3 Other gastrointestinal symptoms

Patil *et al.* (1987) have shown that higher doses of lactitol caused an increase in gastrointestinal side effects following the consumption of one or two boluses per day. In a double-blind, randomised crossover study on 21 normal subjects, Patil *et al.* (1987) determined the laxative threshold of lactitol as being 74 g per 25 h compared to 71 g for sorbitol; however the consumption of up to 40 g lactitol was possible without undesirable side effects. They concluded that the tolerance and laxative threshold of lactitol is similar to that of sorbitol and Patil *et al.* (1987) found that undesirable symptoms appeared at an intake of 40 g per day for both lactitol and sorbitol. Beaugerie *et al.* (1991) showed up to 84% malabsorption of an iso-osmotic solution containing 20 g lactitol without the appearance of intolerance symptoms.

Maltitol is reported to undergo a higher degree of small intestinal hydrolysis than lactitol and, as such, might be expected to cause a lower degree of osmotic diarrhoea and to stimulate fewer of the other symptoms of gastrointestinal intolerance than lactitol. Abraham *et al.* (1981) showed that the low calorie sweetener Marvie®, containing 58% maltitol, was tolerated without any gastrointestinal symptomatology when consumed at between 20 g and 30 g per day although above this level flatulence and abdominal discomfort occurred. Koizumi *et al.* (1983) estimated that a dose of 0.8 g kg^{-1} body weight was the 50% effective dose for laxation and abdominal symptomatology for adult Japanese subjects. Beaugerie *et al.* (1991) indicated 44% malabsorption of 20 g lactitol in iso-osmotic solution without the appearance of intolerance symptoms.

Isomalt contains equimolar quantities of mannitol and sorbitol and may be expected to provoke gastrointestinal intolerance symptoms according to its content of these two monosaccharide alcohols. Fritz *et al.* (1985) found that although there was no reported incidence of diarrhoea when 50 g isomalt was consumed, other intolerance symptoms such as abdominal noise, meteorism and flatulence did occur. The bacterial fermentation of isomalt and its hydrolysates may lead to gastrointestinal reactions if a threshold dose is exceeded and Spengler *et al.* (1987) noted that when healthy subjects were given increasing doses of isomalt in various foods there was a dose-dependent increase in flatulence over a 12 week period. It was concluded that daily doses of 48 g isomalt are tolerated well over an extended period of time. Similarly, daily doses of 24 g of isomalt taken over a 12 week period by Type II diabetics did not stimulate the appearance of intolerance symptoms (Pometta and Trabichet, 1985). Furthermore, Beaugerie *et al.* (1991) showed up to 40% malabsorption of an iso-osmotic solution containing 20 g isomalt without any intolerance.

7.4.4 Biometric studies on polyols in confectionery items

A number of studies have been carried out at the University of Salford to investigate which of the new generation disaccharide alcohols and HGS are best tolerated when incorporated into confectionery items such as chewy sweets and milk chocolate and given to healthy young adult volunteers in properly controlled, randomised, double-blind, longitudinal or crossover field trials. Parameters which have been examined are single and multiple dose administration, acute and chronic consumption, consumption before and after overnight fasting and whether or not a dose-response effect exists. Symptoms which have been examined include flatulence, borborygmi and abdominal pain (colic) as well as stool consistency and frequency.

7.4.4.1 Hydrogenated glucose syrup in chewy sweets. One hundred healthy adults aged 18 to 35 years were invited to consume fruit-flavoured chewy sweets containing either hydrogenated glucose syrup + mannitol (HGS/M) or glucose syrup + sucrose (GS/S) as bulk sweetener. Products were presented as 50 g bags containing 10 × 5 g pieces, each containing a total of 0.4 g bulk sweetener, i.e. a total daily dose of 40 g. Subjects were asked to consume one packet of sweets per day for 14 days, spreading sweet intake *ad libitum* throughout the day. Milk and fruit juice consumption was restricted to 500 cm^3 of each per day and beans were banned. The crossover design was double-blind and randomised with a wash-out period of a week between the two 14-day consumption periods. 91 subjects completed the study. Table 7.4 summarises the gastrointestinal side effects that occurred during the study period. Clearly the HGS/M sweets provoked laxation and this was associated with an increased level of gastrointestinal symptomatology, in particular increased flatus. However, Figure 7.8 shows that the number of subjects passing loose stools on each day of the 14-day HGS/M sweet consumption period decreased and this decrease was highly significant when compared with the period during

Table 7.4 Summary of gastrointestinal symptoms in 100 subjects after daily consumption of sweets containing either 40 g hydrogenated glucose syrup + mannitol (HGS/M) or 40 g sweets containing glucose syrup + sucrose (GS/S)

	(a) HGS/M Sweets	(b) GS/S Sweets
Motion frequency (number per day ± s.e.)	1.41 ± 0.54	1.26 ± 0.48*
Number loose motions in study period (14 days)	414	172**
Number of study days (of 14) when at least 1 loose motion passed	304	137**
Number of subjects with at least 2 loose motions on 1 study day (of 14)	33	9*
Flatus	more	less**
Colic	more	less**

(a) is significantly greater than (b); $^{*}P < 0.01$, $^{**}P < 0.001$

which GS/S were consumed. This trend is apparent whether the criterion of laxation is one or more, two or more or three or more loose motions per day. This decrease in the severity of laxation during the HGS-sweet consumption period indicates that acclimatisation occurred during which the colonic flora presumably adapted to digest HGS more effectively. Consistent with these observations, Kearsley and Birch (1978) and Kearsley *et al.* (1982) showed that the repeated intake of 50 g Lycasin® during a single day on an empty stomach led to diarrhoea but Beaugerie *et al.* (1989) showed that adapted volunteers were symptomless following intake of 69 g per day of Lycasin® during meals. The osmotic effect of polyol or HGS thus depends on the dosage given and the degree of absorption which improves by an adaptive mechanism involving a hormonal effect.

Follow-up studies at Salford were centred on a group of 100 subjects who consumed sweets containing either a moderate or a high level of HGS/polyol (HGS/P) and two sub-groups of 50 from these 100 who consumed either sweets containing a very low level of HGS/P or control sweets. Although all subjects received six packets of sweets they were given in a completely random order. For each consumption test, subjects received product on day 1, adhered to dietary restrictions (as in the previous study) for 24 h, consumed the product on day 2 and were interviewed on day 3 to record their symptomatology. Therefore subjects consumed one bag of sweets on six occasions, each separated by a wash-out period of one week to avoid acclimatisation. The longitudinal study lasted a total of six weeks and the design is shown as a flow chart (see Figure 7.9).

Doses of 50 g, 35 g and 20 g were selected in an attempt to provide HGS/polyol doses above, approximately at and below a hypothetical threshold suggested by the earlier study. Chewy sweets were provided as follows:

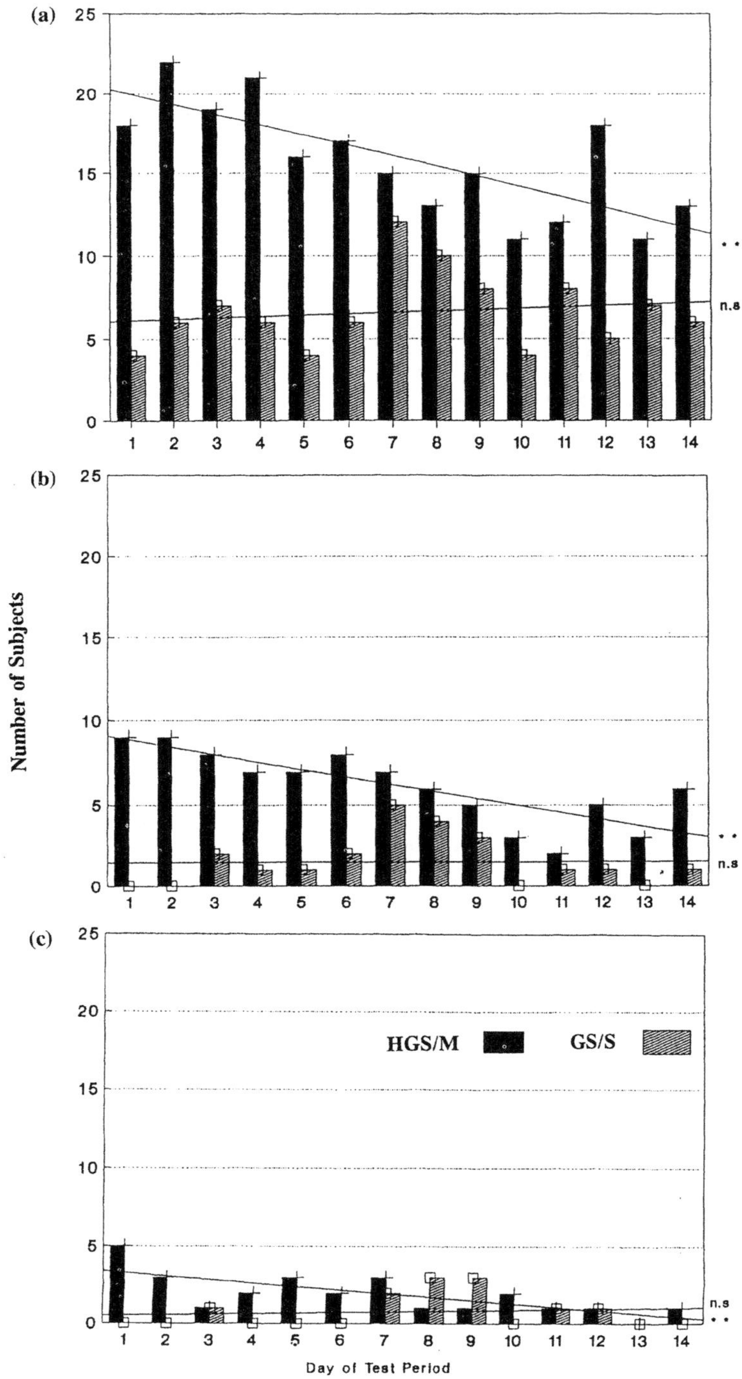

Figure 7.8 Daily incidence of subjects passing loose stools on each of 14 days after daily consumption of sweets containing either 40 g hydrogenated glucose syrup + mannitol (HGS/M) or 40 g glucose syrup + sucrose (GS/S). (a) = 1 or more loose stools per day, (b) = 2 or more loose stools per day, (c) = 3 or more loose stools per day. **is significantly different from n.s. ($P < 0.01$)

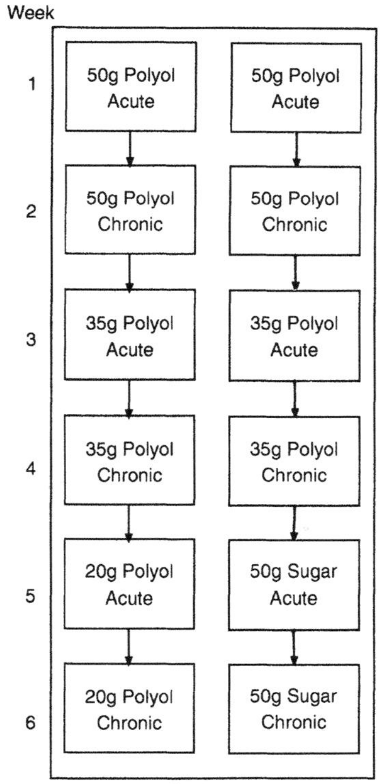

Figure 7.9 Protocol for a double-blind threshold study to evaluate the digestive tolerance of sweets containing either hydrogenated glucose syrup plus polyols (Polyol) at three different levels or controls sweets containing glucose syrup + sucrose (Sugar) when consumed acutely or chronically.

1. Lycasin® solids (83%) + mannitol (12%) + isomalt (5%) = HGS/P
2. Control sweets containing glucose syrup and sucrose = GS/S

To provide subjects with 50, 35 and 20 g HGS/P, bags of sweets contained either 13, 9 or 5 pieces, respectively based on information that each sweet weighed 4.8 g containing 82.1% HGS/P. Hence there were the following sweets available for consumption: HGS/P-50; HGS/P-35; HGS/P-20; GS/S-50.

Consumption was dictated on either an acute or a chronic basis. Acute consumption required that the entire bag of sweets be consumed within 2 h, whereas chronic consumption required that the entire bag be eaten, but that not more than one sweet be eaten in any half hour period during the day.

Table 7.5 summarises the frequency with which motions were passed as well as the percentage of subjects with diarrhoea (defined in this case as the

Table 7.5 Motion frequency and motion consistency in 100 subjects who consumed sweets containing either 50 g, 35 g or 20 g hydrogenated glucose syrup + polyol (HGS/P) or 50 g glucose syrup + sucrose (GS/S), the sweets being consumed in under 2 h (acutely) or at the rate of no more than one sweet every 30 min (chronic)

Product		Subjects (%) not passing motions	Subjects (%) passing motions (mean s.d.)	Mean-post consumption motion frequency				Subjects (%) with diarrhoea
				Total	Loose	Normal	Hard	3 or more loose motions per day
HGS/P-50	Acute	4.6	95.4 (0.93)	1.69	0.55	1.09	0.05	7.1
	Chronic	4.1	95.9 (0.79)	1.51	0.23	1.23	0.05	1.5
HGS/P-35	Acute	5.6	94.4 (0.79)	1.42	0.21	1.16	0.05	1.0
	Chronic	5.1	94.9 (0.72)	1.40	0.11	1.27	0.03	1.0
HGS/P-20	Acute	4.1	95.9 (0.67)	1.37	0.04	1.33	0	0
	Chronic	2.0	98.0 (0.76)	1.39	0.08	1.31	0	0
GS/S-50	Acute	12.2	87.8 (0.69)	1.24	0.08	1.16	0	0
	Chronic	4.1	95.9 (0.53)	1.27	0	1.20	0.06	0

Table 7.6 Gastrointestinal symptomatology in 100 subjects who consumed sweets containing either 50 g, 35 g or 20 g hydrogenated glucose syrup + polyol (HGS/P) or 50 g glucose syrup + sucrose (GS/S), the sweets being consumed in under 2 h (acutely) or at the rate of no more than one sweet every 30 min (chronic)

Symptom		Bloating			Flatulence			Colic		
Product		Mi	Mo	Sv	Mi	Mo	Sv	Mi	Mo	Sv
HGS/P	Acute	23.0	7.1	1.0	33.7	31.1	5.6	17.3	7.7	2.6
	Chronic	19.4	2.6	0.5	36.2	20.4	4.0	12.8	2.0	0.5
HGS/P-35	Acute	16.8	4.1	0	32.6	27.0	1.0	13.8	6.1	1.0
	Chronic	16.3	1.0	0	37.2	16.8	2.5	9.7	2.6	0
HGS/P-20	Acute	8.2	2.0	0	44.9	12.2	0	14.3	0	0
	Chronic	18.4	0	0	40.8	12.2	0	18.4	0	0
GS/S-50	Acute	14.3	0	0	26.5	2.0	0	2.0	0	0
	Chronic	12.2	2.0	0	32.6	4.1	2.0	2.0	0	0

Mi = mild, Mo = moderate, Sv = severe.

passage of loose motions on three or more occasions in the 24 h following the consumption of each product). Table 7.6 gives details of other gastrointestinal symptoms such as bloating, flatulence and colic subjectively ranked using a ten point hedonic scale where 1 to 3 = mild, 4 to 7 = moderate and 8 to 10 = severe. Table 7.7 shows mean symptom scores with the standard error of the mean for all subjects at the various doses taken acutely and chronically. Figure 7.10 is derived from this table and shows the average mean symptom score derived from these data.

Most subjects passed one normal stool during their post-consumption periods but the passing of loose motions was most closely associated with

Table 7.7 Mean symptom scores of 100 subjects who consumed sweets containing either 50 g glucose syrup + sucrose (O) or 20, 35 or 50 g hydrogenated glucose syrup + polyol (20, 35, 50), the sweets being consumed in under 2 h (acute) or at the rate of no more than one sweet every 30 min (chronic). For each symptom the maximum possible score is 10 and the minimum score is 0

		Acute		Chronic	
	Dose	Mean	s.e. mean	Mean	s.e. mean
Stomach ache	0	0.163	0.084	0.143	0.105
	20	0.306	0.143	0.265	0.116
	35	0.413	0.080	0.332	0.069
	50	1.061	0.132	0.342	0.066
Belching	0	0.245	0.095	0.225	0.098
	20	0.204	0.113	0.367	0.119
	35	0.495	0.079	0.291	0.054
	50	0.888	0.124	0.480	0.079
Colic	0	0.061	0.061	0.020	0.020
	20	0.245	0.099	0.306	0.102
	35	0.653	0.116	0.291	0.073
	50	0.959	0.143	0.352	0.077
Wind	0	0.653	0.166	0.898	0.234
	20	1.429	0.253	1.244	0.239
	35	2.189	0.168	1.750	0.161
	50	2.760	0.186	1.974	0.161
Diarrhoea score	0	1.327	0.128	1.184	0.081
	20	1.408	0.096	1.469	0.137
	35	1.592	0.085	1.480	0.069
	50	2.184	0.130	1.674	0.090
Average	0	0.490	0.071	0.494	0.068
	20	0.718	0.087	0.727	0.093
	35	0.168	0.066	0.829	0.054
	50	1.570	0.099	0.964	0.058

consumption of the HGS/P-50 and HGS/P-35 sweets. This association was enhanced further by acute consumption. There was a significant increase ($P < 0.01$) in the occurrence of diarrhoea with increasing dose of HGS/P and this increase was faster than linear for acute consumption but not for chronic consumption. There was also a significant increase in flatulence during both acute and chronic consumption of product as the dose increased. In this case the increased incidence of flatulence was linear in both cases, though greater during acute consumption. Although there was an increase in colic, reporting of this symptom was low for all products irrespective of dose or consumption regime.

A similar study involved using otherwise identical sweets but in this case containing 83% Lycasin solids plus 22% isomalt with numbers of sweets being controlled to give 50 g and 20 g doses. The same trends were apparent in relation to acute as compared to chronic consumption and also in relation to increasing polyol dose. However, more people provided unprompted information on gastrointestinal symptoms when consuming

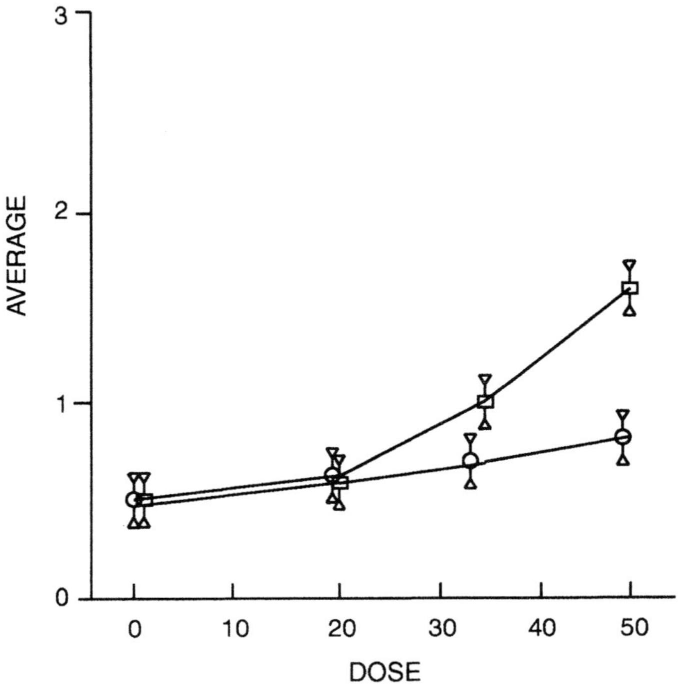

Figure 7.10 Mean symptom average score for 100 subjects consuming either hydrogenated glucose syrup plus polyols at three different levels or control sweets containing glucose syrup + sucrose when consumed acutely (–□–) or chronically (–○–) [±2 standard errors of the mean (–△–)].

HGS/P-50 rather than HGS/I-50, but there was no difference in laxativity between the two products.

It may be concluded that consumption of 35 g HGS/P is associated with a tendency towards increased motion frequency and also towards looseness of stools. This tendency is enhanced somewhat if the 35 g dose was ingested in under 2 h or if the dose was increased to 50 g. It would appear that 35 g HGS/P-35 is at, or slightly above the level tolerated by most people but acute consumption tended to enhance the appearance of symptoms in susceptible individuals. Ingestion of 50 g HGS/P exceeded this level of tolerance and a small but increased number of individuals reported intolerance symptoms.

7.4.4.2 Polyols in chocolate. Maltitol is a widely used sweetening agent because of its physicochemical properties, natural sweetening power (80% that of sucose) and low cariogenicity. Sucrose is generally considered to be cariogenic and standard milk chocolate usually contains sucrose in the range of 40 to 45% as an added ingredient. However, replacement of sucrose in chocolate by polyols such as maltitol is possible taking account of a variety of technological properties such as thermal stability, hygroscopicity, sweetness and water content (Zumbé *et al.*, 1994). High purity, crystalline D-maltitol such as Maltisorb® is an excellent substitute for sucrose in sugarless chocolate for technological and organoleptic reasons (Le Bot, 1993; Oliger, 1994; Rapaille *et al.*, 1994; Sicard and Le Bot, 1994).

The gastrointestinal symptoms resulting from eating milk chocolate containing maltitol have been compared with those occurring in the same individuals after eating normal, sucrose-containing chocolate (Storey *et al.*, submitted for publication). In a double-blind crossover study, 20 healthy volunteers (9 males and 11 females between the ages of 18 and 24 years having body mass indices (mean/s.d. of: male 25.1 (3.9) and female 22.0 (1.2)) ingested 100 g bars of milk chocolate which were identical except for their added carbohydrate ingredient. Each 100 g bar contained 40 g bulk sweetener; one product contained sucrose (S), one maltitol (M) and the other a mixture of sucrose and maltitol (SM, 10:30 w/w). Subjects were tested twice with each product given in randomised order determined by a latin square design applied to each leg of the crossover study which related to consumption in fasting and non-fasting conditions. One group consumed S, SM and M during weeks 1 to 3 while fasting and again during weeks 4 to 6 while not fasting. The other crossover group of ten subjects first consumed the products while not fasting and later while fasting. Consumption periods were separated by wash-out periods of 1 week. For 12 h before the test day, the 24 h of the test day and for 12 h thereafter subjects restricted their milk and fruit juice intakes to 300 cm^3 each per 24 h and refrained totally from drinking alcohol. Fasting subjects took no food or liquid after 22.00 the night before the chocolate consumption day and consumed the test product as breakfast, being allowed to drink up to 200 cm^3 of liquid to ease consumption; a normal meal pattern was resumed from noon on the day of consumption. Non-fasting subjects consumed products within 30 min of the completion of their normal breakfast. Both groups were instructed to eat the entire 100 g chocolate bar within 30 min.

The number of toilet visits made during the 24 h post-consumption was recorded and the consistency of faeces passed on each occasion classified as either watery, normal or hard. Gastrointestinal symptoms were recorded for the 24 h period following consumption as the separate sensations of borborygmi, colic and flatus by ranking them on a hedonic scale where 0 = normal function, 1 = mild symptoms, 2 = moderate symptoms and 3 = severe symptoms. Table 7.8 gives details of intolerance symptoms and Figure 7.11 shows the mean symptom score derived from these data.

Storey *et al.* (submitted for publication) discovered that there was no difference in symptomatology between fasting and non-fasting subjects ($P > 0.05$) nor according to whether maltitol-containing chocolate was

Table 7.8 Intolerance symptoms after ingestion of 100 g chocolate containing either 40 g sucrose, 30 g maltitol (+ 10 g sucrose), or 40 g maltitol irrespective of fasting and non-fasting (from Storey *et al.* submitted for publication) (number of subjects reporting symptoms from groups of forty)

Symptom	Sucrose	30 g Maltitol	40 g Maltitol
Borborygmi	5	8	11*
Abdominal colic	4	2	7
Flatus	11	16	23**
Loose faeces	6	4	9
Mean number of toilet visits (SEM)	1.25 (0.01)	1.38 (0.09)	1.4 (0.10)

*Significant increase in symptom with 40 g maltitol compared to sucrose ($P < 0.05$).
**Significant increase in symptom with 40 g maltitol compared to sucrose ($P < 0.01$).

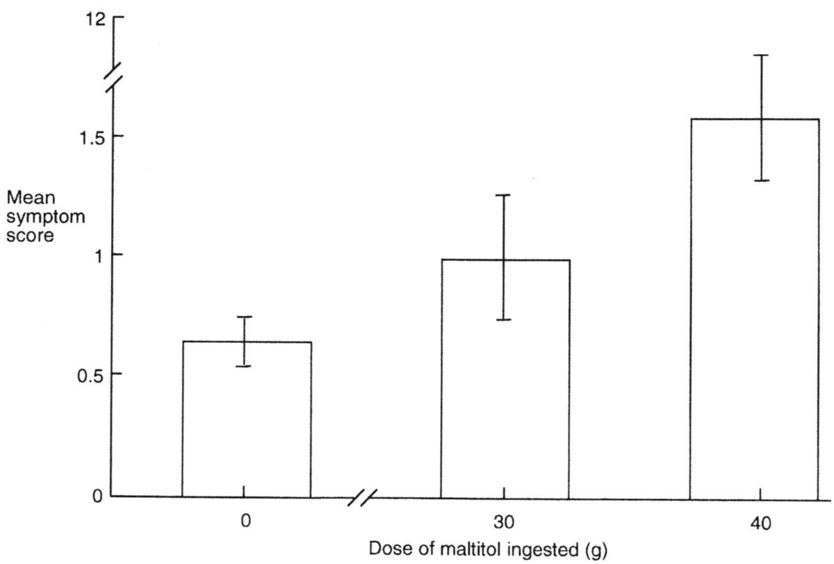

Figure 7.11 Mean symptom score for 20 subjects after ingestion of 100 g chocolate containing either 0, 30 or 40 g maltitol. (The maximum score possible was 12 and a score of 1 would mean that on average subjects suffered 1 mild symptom on 1 occasion) (after Koutsou *et al.*, 1995).

eaten before or after normal chocolate ($P > 0.05$). Eating 30 g maltitol in chocolate caused no increase in gastrointestinal symptoms but ingestion of 40 g increased mild borborygms ($P < 0.05$) and mild flatus ($P < 0.01$). Ingestion of 30 g ($P = 0.55$) or 40 g ($P = 0.24$) maltitol did not cause any significant increase in laxation; nine subjects passed loose faeces on one occasion after eating 40 g maltitol but this was not significantly greater than after consumption of 30 g maltitol (four subjects) or 40 g sucrose (six subjects). Colic and loose faeces, potentially more distressing symptoms, were uncommon even with the higher dose of 40 g maltitol and the incidence was not significantly higher than after consumption of the 30 g maltitol or sucrose-containing chocolate. More subjects reported multiple symptomatology following consumption of 40 g maltitol (18) and 30 g maltitol (7) compared to sucrose consumption (3; $P < 0.01$) suggesting that certain individuals are more susceptible than others to incompletely hydrolysed maltitol and unabsorbed sorbitol entering the large intestine. This may be due in part to intersubject variations in symptom perception and description, but there is little doubt that some individuals appear to be more susceptible than others to the consequences of incompletely hydrolysed maltitol entering the large intestine along with unabsorbed sorbitol (Dahlqvist and Telenius, 1965; Nilsson and Jagerstad, 1987). These results are consistent with those of Abraham *et al.* (1981) and Beaugerie *et al.* (1991) who demonstrated that a 20 to 30 g dose of maltitol syrup could be well tolerated without undesirable symptoms.

These observations of Storey *et al.* (submitted for publication) that an acute oral intake of 30 g crystalline maltitol in milk chocolate resulted in no significant adverse symptoms and that there was no significant laxative effect when ingested at either 30 g or 40 g per day have major implications for manufacturers of confectionery products. The symptomatology arising from maltitol consumption was not only very low but was also broadly similar in fasting and non-fasting subjects, a consideration of some importance in relation to the propensity of consumers to eat confectionery items as intermeal snacks. It seems that a partitioning of unhydrolysed maltitol and unabsorbed sorbitol may have occurred in both fasting and non-fasting subjects leading to a similar gradual delivery of such molecules to the large intestine.

7.4.4.3 Comparative studies of polyols in chocolate. Lactitol and isomalt are two other disaccharide alcohols which have potential as bulk sweeteners for incorporation into confectionery products such as chocolate. There is a considerable literature relating to the appearance of intolerance symptoms following the consumption of these polyols in solution and various other formulations and also information exists relating to possible threshold values below which there is negligible symptomatology. However very few published studies have investigated the ingestion of either lactitol or isomalt incorporated into chocolate. Livesey *et al.* (1993) found that consumption of lactitol-containing chocolate, isomalt-containing chocolate and polydextrose-containing chocolate resulted in 112%, 73% and 11% increases in breath H_2 respectively compared to sucrose-containing chocolate, this being indicative of a high degree of intestinal fermentation. Broad support was given to these observations by the work of Lee *et al.* (1994) who found that ingestion of 31.5 g doses of either isomalt or sorbitol incorporated into milk chocolate led to a sharp post-consumption increase in breath H_2 excretion compared with that after consumption of an otherwise identical sucrose-containing chocolate. Total breath H_2 excretion after eating sorbitol chocolate was significantly greater than after eating isomalt chocolate ($P < 0.01$). Zumbé and Brinkworth (1992) found that consumption of isomalt-containing chocolate was associated with an increase in motion frequency and flatulence in both normal and Type II diabetics.

Koutsou *et al.* (1995) have recently compared chocolate containing either isomalt, lactitol or maltitol, three disaccharide alcohols which are increasingly being used as bulk sweeteners in milk chocolate and other chocolate-containing confectionery items such as snack bars. They assessed the effects of ingesting different polyols incorporated into standard milk chocolate by assessing the incidence and severity of gastrointestinal symptoms and comparing them with those occurring in the same individuals after eating normal sucrose-containing chocolate. In a double-blind crossover study 59 healthy volunteers aged 18 to 24 years ingested 100 g chocolate containing 40 g bulk sweetener as either sucrose, lactitol, isomalt or maltitol or mixtures of sucrose and polyol (10:30 w/w) to give 30 g doses. Table 7.9 gives details of the intolerance symptoms and shows that 30 g and 40 g lactitol caused a significant increase in flatulence, borborygmi, colic, stool frequency and looseness compared to sucrose ($P < 0.01$). 40 g isomalt increased all symptoms including mild laxation ($P < 0.01$) but, unlike lactitol, none was rated as severe. Reduction of isomalt to 30 g was marked by increased tolerance with no increase in stool frequency compared to sucrose ($P = 0.35$). and evidence of only mild borborygmi ($P < 0.01$) and mild flatulence, colic and laxation ($P < 0.05$) 40 g maltitol caused even less intolerance than 40 g isomalt and this was reduced further by a reduction to 30 g. Maltitol had no laxative effect when consumed at either 30 g ($P = 0.32$) or 40 g ($P = 0.13$). All these comparisons were made between polyol-containing chocolates and a standard sucrose-containing chocolate bar.

Table 7.10 compares the symptomatology produced by chocolate containing maltitol, isomalt and lactitol. Very few significant differences were evident between products, most being between lactitol and the other two polyols. However chocolate containing 40 g isomalt produced somewhat looser stools than that containing 40 g maltitol, whereas 40 g maltitol provoked greater flatus than 40 g isomalt. There were no differences in symptomatology between chocolate containing 30 g isomalt and 30 g maltitol.

The relatively low levels of gastrointestinal symptomatology recorded in the studies of Lee *et al.* (1994), Koutsou *et al.* (1995) and Storey *et al.* (submitted for publication) may relate to the fact that in these studies the polyols were eaten in chocolate rather than in solution or in other non-fat-containing confectionery products such as chewy or boiled sweets. Milk

Table 7.9 Intolerance symptoms after ingestion of 100 g chocolate containing either 40 g sucrose, 30 g maltitol 30 g isomalt or 30 g lactitol (+ 10 g sucrose in each case), or 40 g maltitol, 40 g isomalt or 40 g lactitol (from Koutsou *et al.*, 1995; number of subjects reporting symptoms from groups of 59)

	Bulk sweetener						
Symptom	Sucrose 40 g	Maltitol 30 g	Isomalt 30 g	Lactitol 30 g	Maltitol 40 g	Isomalt 40 g	Lactitol 40 g
Borborygmi							
total	8	16	26[a]	26[a]	28[a]	29[a]	34[a]
mild	7	10	13	17	15	17	21
moderate	1	6	6	10	10	9	9
severe	0	0	7	2	3	3	4
Colic							
total	3	10	14[b]	17[a]	15[a]	17[a]	26[a]
mild	3	10	10	12	8	7	16
moderate	0	0	4	4	6	8	9
severe	0	0	0	1	1	2	1
Flatus							
total	16	33[a]	31[b]	42[a]	40[a]	43[a]	51[a]
mild	10	20	22	21	16	26	20
moderate	6	12	7	17	16	15	22
severe	0	1	2	4	8	2	9
Loose faeces							
total	6	8	16[b]	21[a]	12	23[a]	32[a]
1	3	6	9	13	10	14	14
2	2	0	5	3	0	5	8
3+	1	2	2	5	2	4	10

[a] Significant increase in symptom compared with sucrose product ($P < 0.01$).
[b] Significant increase in symptom compared with sucrose product ($P < 0.05$)

chocolate typically contains approximately 30% fat which, in any digestible form in the presence of bile and pancreatic lipase, is the most powerful agent in the retardation of gastric emptying. Fat digestion products, in particular fatty acids, stimulate the production of cholecystokinin (CCK) by the mucosa of the duodenum and jejunum and CCK then reduces the amplitude of gastric peristalsis which retards gastric emptying. Such an effect is likely to prolong the period of polyol delivery to the small intestine thus ensuring its presentation to the brush border enzymes at a favourable rate.

Nevertheless, the study of Koutsou *et al.* (1995) confirmed that there are differences between the disaccharide alcohols in terms of their ability to provoke gastrointestinal symptoms. It is likely that this is related primarily to the relative degree of small intestinal hydrolysis and absorption which the polyols undergo. Maltitol undergoes about 10% hydrolysis compared with maltose to yield glucose which is actively absorbed, sorbitol which is

Table 7.10 χ^2 values from statistical analysis of symptomatology using 2 × 2 transition table analysis and formula of McNemar. (Values in parentheses are those from 4 × 4 transition table analysis for severity of symptomatology). All significant differences (except 40 g maltitol versus 30 g isomalt* for symptoms of flatus) are increases in symptomatology upon consumption of polyol mentioned second compared to that mentioned first

	30 g				40 g			
	Borborygmi	Colic	Flatus	Loose stools	Borborygmi	Colic	Flatus	Loose stools
Maltitol versus isomalt (30 g)	3.682 (10.77)	0.563 (4.077)	0.409 (5.259)	2.45 (8.286)	5.76[b] (8.17)	2.11 (10.00)	3.375 (7.54)	7.84[a] (11.56)
Maltitol versus isomalt (40 g)	0.346 (4.50)	0.0625 (3.13)	4.50[b]* (15.28[b])	0.00 (5.50)	0.00 (1.78)	0.42 (2.11)	0.211 (7.867)	5.5[b] (10.77)
Isomalt versus lactitol (30 g)	0.045 (5.286)	0.45 (2.862)	6.722[a] (14.051[b])	0.563 (4.13)	2.227 (8.131)	7.042[a] (8.571)	16.409[a] (24.10[a])	8.654[a] (15.857[b])
Isomalt versus lactitol (30 g)	1.389 (5.901)	0.0833 (3.333)	0.190 (2.095)	1.389 2.933	0.941 (6.685)	3.048 (9.269)	4.923[b] (12.818[b])	2.37 (11.222)
Maltitol versus lactitol (30 g)	5.76[b] (8.384)	1.895 (5.60)	3.368 (8.104)	6.857[a] (10.57)	10.32[a] (15.778[b])	9.375[a] (14.765[b])	13.136[a] (21.122[a])	14.694[a] (19.919[a])
Maltitol versus lactitol (40 g)	3.368 (6.69)	0.063 (4.60)	0.125 (5.086)	3.368 (6.692)	1.389 (6.78)	4.762[b] (8.934)	7.692[a] (10.40)	15.042[a] (21.935[a])

[a] Significant difference at 1% level (2 × 2 tabulated $\chi^2 = 6.64$: 4 × 4 tabulated $\chi^2 = 16.81$)
[b] Significant difference at 5% level (2 × 2 tabulated $\chi^2 = 3.84$: 4 × 4 tabulated $\chi^2 = 12.59$)

passively absorbed (by facilitated diffusion and solvent drag) and residual polyol which enters the large intestine along with unabsorbed sorbitol both of which are then fermented by bacteria, primarily in the proximal colon. Lactitol undergoes less than 3% hydrolysis compared with lactose (Grimble *et al.*, 1988; Metzger *et al.*, 1988; Nilsson and Jagerstad, 1987; Wursch *et al.*, 1989) leaving residual polyol which presumably stimulates laxation because of the high osmotic load in the ileum. Like lactitol, isomalt undergoes marginal hydrolysis in the proximal small intestine (Grupp and Siebert, 1978; Nilsson and Jagerstad, 1987) but the levels of gastrointestinal symptomatology recorded by Koutsou *et al.* (1995) presumably reflect not only differences in the extent of hydrolysis but also differences in the dynamics of fermentation of the mixture of residual polyol and polyol hydrolysate by the colonic microflora (Livesey *et al.*, 1993) as well as the better absorption of isomalt in the small intestine (Beaugerie *et al.*, 1991; Wursch *et al.*, 1989). Notwithstanding these differences, Koutsou *et al.* (1995) demonstrated that symptomatology was always dose related and even with lactitol many intolerance symptoms decreased in incidence and severity when the level of ingestion was decreased.

7.4.4.4 Breath hydrogen excretion. The normal human digestive tract contains about 150 cm^3 of gas of which approximately 100 cm^3 is in the colon. When unabsorbed carbohydrate enters the colon and is fermented, the total amount and proportion of gases produced depends upon the nature of the fermented residues, but in any case much of the gas is absorbed and excreted across the lung epithelium although some is also consumed by other bacteria. Breath H_2 testing has been used by many workers to provide information on the extent and duration of fermentation of unabsorbed dietary components (Bond and Levitt, 1972; Flourié *et al.*, 1988a and b; Fritz *et al.*, 1985; Hyams, 1983; Lee *et al.*, 1994; Oku *et al.*, 1991; Wursch *et al.*, 1989). However the quantitative aspects of breath H_2 excretion must be considered with caution especially when comparisons are to be made because of large inter- and intra-individual differences between individuals (Rumessen *et al.*, 1990) and because of differences in the stoichiometry of H_2 production following the consumption of different polyols (Livesey *et al.*, 1993). Hammer (1993) demonstrated that colonic H_2 absorption is highly effective at low colonic hydrogen accumulation rates (i.e. less than 76 cm^3 H_2/6 h) but not so at higher concentrations when absorption can decrease to as little as 20%. Nevertheless, inter-individual differences in breath H_2 excretion after the ingestion of a standard dose of carbohydrate (e.g. 12.5 g lactulose) are probably due to differences in bacterial net gas production rather than differences in gas absorption. That is, incomplete colonic gas absorption is the consequence, and not the cause, of high colonic gas accumulation rates.

Notwithstanding these difficulties in the interpretation of breath H_2 excretion, the technique has a certain attractiveness because it is non-invasive and does not involve the use of radiolabelled molecules. Furthermore, other available techniques suffer from many of the same limitations. Storey *et al.* (submitted for publication) have used the breath hydrogen test to assess the colonic fermentation of polyols ingested in milk chocolate rather than in liquid form and have shown that the ingestion of milk chocolate containing maltitol leads to an elevated breath H_2 response compared with that following the consumption of milk chocolate containing only sucrose, a difference which is significant at the 1% level and is indicative of a greater degree of intestinal gaseousness due to bacterial fermentation of maltitol and sorbitol. Non-adapted, fasting subjects were tested with chocolate product in a randomised order with test periods separated by 1 week. Test products were consumed in 30 min immediately prior to breath testing and the positive control was a 20 ml dose of lactulose syrup. A sharp rise in breath H_2 concentrations after polyol consumption in chocolate marked the beginning of colonic fermentation, effectively the orocaecal transit time, being 115.5 min. ($\pm$ 18.7) for milk chocolate containing 30 g maltitol and 87.0 min ($\pm$ 14.3) for a similar product containing 40 g maltitol. Ingestion of 40 g maltitol in chocolate led to a greater total H_2 excretion compared with that after eating chocolate containing either 30 g maltitol ($P < 0.05$) or sucrose ($P < 0.01$). In a similar study Lee *et al.* (1994) showed that breath H_2 excretion after consumption of chocolate containing either sorbitol or maltitol was significantly higher than after consumption of chocolate containing sucrose ($P < 0.01$). The dose-related breath H_2 response observed by Storey *et al.* (submitted for publication) was consistent with a lower reported symptomatology following ingestion of 30 g maltitol compared with 40 g maltitol. The total breath H_2 excretion after lactulose was 183.1 ppm $\pm$ 40.8 and, taking this as 100%, it was calculated that on average, 34.7% (10.4 $\pm$ 1.4) of a 30 g ingested dose and 49.2% (19.7 g $\pm$ 4.0) of a 40 g ingested dose of maltitol was fermented in the colon. An estimated maltitol hydrolysis and constituent sugar absorption of ca. 65% of a 30 g dose and ca. 50% of a 40 g dose is in broad agreement with the 50% malabsorption of maltitol reported by Wursch *et al.* (1989) and Beaugerie *et al.* (1991).

The dynamics of the gastointestinal response to the ingestion of hydrogenated glucose syrups and polyols depend upon many factors including (i) the extent of digestion of the native polyol, (ii) the extent of absorption of the products of hydrolysis, (iii) the inherent fermentability of the residual polyol and its hydrolysis products, (iv) the fermentative capacity of the particular colon under consideration, (v) the extent of colonic absorption of fermentation products, (vi) the osmotic properties of the polyol/hydrolysis products mixture, (vii) the 'sensitivity' of the large bowel *muscularis externa*. One aspect of the work of Lee *et al.* (1994) and

Storey *et al.* (submitted for publication) concerning the ingestion of polyols incorporated into milk chocolate was to attempt to correlate the breath hydrogen response (BHR) with the incidence of gastrointestinal symptomatology in non-intubated subjects thus taking all the above considerations into account. Livesey *et al.* (1993) pointed out that the interpretation of the BHR is confounded by the differing stoichiometries for hydrogen production, by interaction between substrates and by uncertainty concerning the extent to which small intestinal hydrolysis yields species with a stoichiometry that differs from the parent substrate. Nevertheless, although the interpretation of BHR is obviously complex and must be made with circumspection, there seems to be little doubt that, taken together with information relating to the incidence of intolerance symptoms in the individuals providing the breath samples, it does give a useful insight into the events occurring in the intact gut concerning the degree of digestion and fermentation of ingested polyol.

Livesey *et al.* (1993) refer to the low production of H_2 from isomalt compared to other polyols and polyol mixtures after *in vitro* faecal fermentation. We have also performed such studies and one relevant data set is presented in Figure 7.12 which confirms that crystalline maltitol as

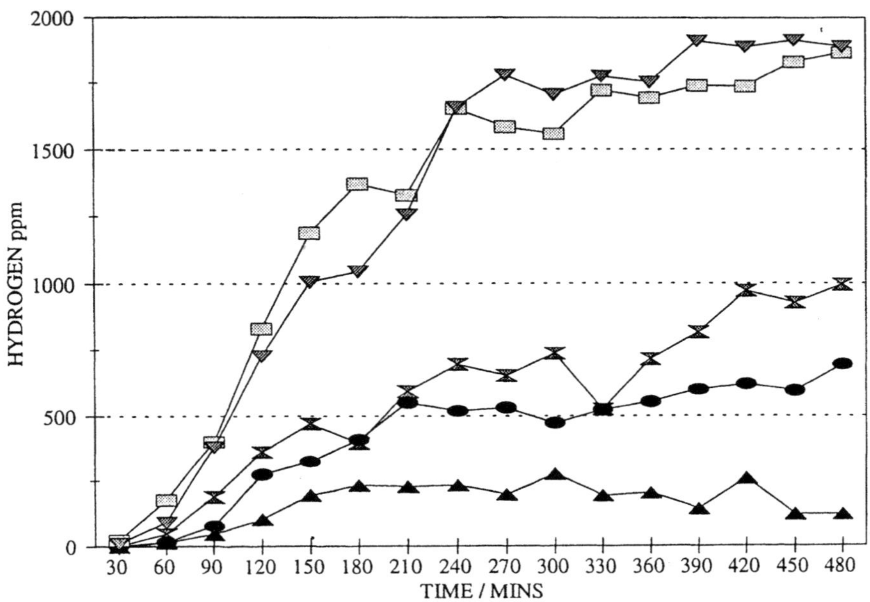

Figure 7.12 Hydrogen concentrations (ppm) over an 8 h period from *in vitro* faecal fermentations of lactitol, sorbitol, isomalt and maltitol in 100 cm^3 sterile growth medium as well as a control containing no polyol. Values are the mean of 14 replicate determinations. Sorbitol/lactitol were significantly different from isomalt/maltitol ($P < 0.05$). –▲–, control; –⧗–, maltitol; –●–, isomalt; –▩–, sorbitol; –▼–, lactitol.

well as isomalt appear to be inherently less fermentable than some other sugar alcohols such as sorbitol and lactitol. It must be borne in mind, however, that the confines of an inert *in vitro* fermentation vessel cannot replicate exactly the conditions within a human colon and also, as yet, there is no way of obtaining a standardised faecal inoculum. *In vitro* faecal fermentation is a useful technique, but much remains to be defined in terms of its suitability to predict with confidence what might happen in the colon of a person ingesting polyol. Like breath hydrogen monitoring, it should be used with caution along with data from consumption studies to explain the gastrointestinal response to polyols and mixtures containing polyols (Storey *et al.*, 1994).

7.5 Use of low digestible carbohydrates in confectionery products

7.5.1 Legal issues

7.5.1.1 Status of polyols as foods or food additives. The food industry uses polyols as bulk sweeteners to produce excellent tasting products with special physiological properties, a lower caloric value compared to sucrose-containing products, a hypocariogenic effect and a low blood glucose response enabling their use by diabetics. In this respect polyols may be considered as 'functional foods' (Goldberg, 1994). Table 7.11 shows a

Table 7.11 Functional food ingredients (from Elliot, 1994)

Functional ingredient	Potential functional benefit
Antioxidant vitamins	Prevent CHD, cancer, cataracts
Dietary fibre	Lowers serum cholesterol
	Prevents colon and breast cancers
Oligosaccharides	Produce lower metabolizable energy than sucrose
	No effect on insulin secretion
	Improve colon microflora
	Prevent dental caries
Minerals	
calcium	Reduction of osteoporosis
magnesium	Prevents heart and skeletal muscle dysfunction, brain dysfunction
zinc	Prevents loss of immune function
Omega-3 fatty acids	Reduce incidence of CHD
Phytochemicals	Reduce serum cholesterol
	Prevent cancer
Sugar alcohols	**Non-caloric sweeteners**
	Prevent dental caries
Lactic acid bacteria	Maintain healthy intestinal flora
	Improve lactose intolerance
	Prevent diarrhoea
	Lower cholesterol
	Boost immune system

classification of functional food ingredients and gives details of potential functional benefits. Sugar alcohols are stated as being 'non-caloric sweeteners' which 'prevent dental caries'. The European Union has made some progress in recognising functional foods as a new category of foods and systems have been put in place that grant exclusivity for innovative new products based on proprietary research (Labell, 1993).

However, from a legal point of view polyols are still classified as food additives under the legislation of the EU and that of many individual member countries. This is astonishing because this classification seems not to comply with the legal definition of a food additive. Table 7.12 gives the definition of a food additive within the EU (1989a) and *Codex Alimentarius* (1979) both of which require the addition for technological purpose. Further, the additive should not be consumed as a food in itself nor should it be present as a characteristic ingredient of food. These definitions clearly do not apply to polyols which are found naturally in considerable amounts as a major constituent of many foodstuffs (Tables 7.1 and 7.2). Polyols are

Table 7.12 Definition of 'food additive' and 'food' within the *Codex Alimentarius* (1979) and the European Union (1989a)

	Codex Alimentarius	European Union
Food additive	Any substance not normally consumed as food by itself and not normally used as a typical ingredient of the food, whether or not it has nutritive value, the intentional addition of which to food for a technological (including organoleptic) purpose in the manufacture, processing, preparation, treatment, packing, packaging, transport or holding of such food results, or may be reasonably expected to result (directly or indirectly) in it or its byproducts becoming a component of or otherwise affecting the characteristics of such foods. The term does not include 'contaminants' or substances added to food for maintaining or improving nutritional qualities.	Any substance not normally consumed as a food in itself and not normally used as a characteristic ingredient of food whether or not it has nutritive value, the intentional addition of which to food for a technological purpose in the manufacture, processing, preparation, treatment, transport or storage of such food results, or may be reasonably expected to result, in it or its byproducts becoming directly or indirectly a component of such foods.
Food	Foodstuff: Any substance, whether processed, semi-processed or raw, which is intended for human consumption, and includes drink, chewing gum and any substance which has been used in the manufacture, preparation or treatment of 'food' but does not include cosmetics or tobacco or substances used only as drugs.	No definition for food existing

also characteristic ingredients of 'sugar-free' and 'no-sugar-added' products and their intentional addition is to take advantage of their physiological properties (Fondu and Gerrard, 1991; Stein, 1992). Furthermore, the Joint Expert Committee on Food Additives (JEFCA, 1983–1986; WHO, 1983, 1990) and the Scientific Committee on Food of the EU (SCoFEU, 1984) both concluded in their safety evaluations that polyols are safe for human consumption. JEFCA concluded that an 'ADI' need not be specified for these macronutrients.

On this basis, the authors of this paper propose that polyols be reclassified as (functional) foods and not as food additives.

7.5.1.2 Laxative effect of polyols. The possible laxative effect of polyol consumption has not been considered by JEFCA and SCoFEU as being a matter of safety in terms of toxicology. JEFCA noted that polyols (and other substances such as some gums, modified starches and modified cellulose) may have a laxative effect under certain circumstances (WHO, 1990). In its sweetener report (Commission of the European Communities, 1987) SCoFEU noted that, "Laxation may be observed at high doses. Consumption of the order of 20 g per person per day of polyols is unlikely to cause undesirable laxative symptoms. The level for individual polyols ingested singly is higher in many cases".

On this basis the health authorities in a number of countries were of the opinion that the ultimate consumers should be informed about the possible laxative effect of polyols which might occur under certain circumstances. However, the conditions leading to a warning label vary from country to country (Table 7.13). In some countries, the warning is required if a defined percentage of polyols is present in the final product. In others a warning has to be given whenever a polyol is used. Examples of various warning labels are given in Table 7.14. Nevertheless, it is worthy of note that in several countries, e.g. UK, Canada, Finland, New Zealand, a statement on laxativity is not required by law. Worldwide, there is a large variety of conditions and required statements regarding laxation warning.

Table 7.13 Review of conditions making the use of a warning statement on laxativity mandatory in various countries

Country	Conditions
Australia	If a polyol is used
Belgium	Depending on product group and maximum use level
European Union	If more than 10% polyols are used in the final product (DRAFT)
France	If polyols are used
Germany	If more than 10% polyols are used in the final product
Luxembourg	If more than 10% polyols are used in the final product
Switzerland	If more than 10% polyols are used in the final product

Table 7.14 Examples of warning statements on laxativity

Country	Statement on laxativity
Australia	Excess consumption may have a laxative effect
Belgium	Do not eat more than 50 g daily
European Union	Excessive consumption may have laxative effects
France	Daily excess consumption may have a slight laxative effect Do not give to children younger than three years
Germany	Excess consumption may have a laxative effect
Luxembourg	Excess consumption may have a laxative effect
Switzerland	Can have a laxative effect

The latest proposal concerning a warning label for polyols within the EU requires the warning statement "excessive consumption may have a laxative effect" if more than 10% of polyols are present in the final product (European Union, 1994b). Questions may be asked as to whether such warnings have any meaning at all and whether they may even mislead the consumer.

A consequence of the 1994 draft EU directive would be that nearly all products where polyols are used for sweetening purposes would have to bear a statement on laxativity. In practice, it is known that problems of intestinal discomfort following polyol ingestion are uncommon. In Switzerland, the world's largest market for sugar-free products, a consumer food tolerance survey was carried out on a representative sample of 1000 Swiss people to analyse the tolerance of sugar-free products, mainly sugar-free confectionery (SRU, 1992). Only 0.5% spontaneously claimed to be sensitive or 'allergic' to sugar-free products and the proposed warning would, therefore, have been meaningless for all but a tiny fraction of this study population. Warning labels appear to be justified only in cases where a significant risk of overconsumption exists caused by the intake of large amounts of polyol in a short period of time; otherwise, a situation could arise with the proposed warning, that consumers might become uncertain regarding the safety of the product without justification. Seventy percent of the world's adult population, including 30% in certain European countries, are lactose intolerant (Bayless *et al.*, 1971; Johnson, 1981; Paige *et al.*, 1975) and yet there is no warning label on milk or milk products! There seems little doubt that consumers will avoid foods which cause them to suffer adversely, but will ignore warning statements which are, for them, meaningless.

The authors believe that when consumers become familiar with polyol-containing products, warning labels will become redundant. We propose that if food law requires a warning label on final products containing polyols, it should be limited to a fixed period of time (e.g. 3 years), after which consumers (especially those who are more sensitive than others) would have become familiar with the products and their properties. In the USA such a time period was recently introduced when changing the standard of identity for frozen desserts. FDA required that all non-nutritive sweeteners used in food should be declared as part of the food (Food and Drug Administration, 1994). However, this requirement will be terminated after a period of 3 years, which FDA believes is adequate for people to become aware of the change.

7.5.1.3 Recommendations regarding laxativity warnings. Scientific data on the gastrointestinal tolerance of polyols reveal large inter- and intra-individual variations, as a consequence of which, laxativity warnings on all polyol-containing products cannot be justified. Data indicate that disaccharide alcohols such as isomalt and maltitol are tolerated somewhat better than are monosaccharide alcohols such as sorbitol and xylitol. The following recommendations are suggested:

- A suitable warning might be: 'Excess consumption may have laxative effects'.
- Warnings should be limited to products where the expected intake pattern indicates that overconsumption might occur (e.g. > 20 g as a single dose in under 1 h).
- Such a warning should apply to:
  - (i) foodstuffs containing over 10% of added polyols,
  - (ii) products containing over 10 g monosaccharide alcohols per unit product,
  - (iii) products containing over 20 g disaccharide alcohols per unit product (disaccharide alcohols to include up to 10% monosaccharide alcohols).
- Where the criteria above are not exceeded, warning labels on laxativity should be voluntary and selective.
- In the case of chewing gums and certain candies, for which the portion size contains less than a total amount of polyol as indicated in (ii) or (iii) above, a laxative warning does not make sense and should not be required.
- If laxative warnings are required by law, they should be restricted to a certain period of time (e.g. 3 years).

7.5.1.4 'Suitable for diabetics' claims. As with most other people, those suffering from diabetes sometimes like to eat sweets or snacks as part of their diet. According to the nutritional recommendations for diabetics, carbohydrates with a low glycaemic index are preferred and total energy intake should always be restricted to reduce or avoid overweight and obesity (European Association for the Study of Diabetes, 1988). Although these recommendations focus on the total daily diet, they also appear to be valid for single foods.

Polyol-containing foods can increase the quality of life for diabetics by allowing them to introduce convenient, appealing and excellent-tasting foods into their diet without provoking adverse effects on their blood glucose levels. The glycaemic index of polyol-containing foods is low or zero and they have a reduced energy value compared to similar foods containing sucrose. Of course such foods of low glycaemic index are generally good for the population at large as well as diabetics. In a number of countries, polyol-containing foods can currently be labelled as 'suitable for diabetics'. However, the future for such products is uncertain within the EU at the moment. A specific directive on foods for diabetics was foreseen within the EU framework on particular nutritional uses (European Union, 1989) but the situation changed during the Community meeting in Edinburgh in 1992. It was decided to restrict the Community's legislative activity to what is strictly necessary for the functioning of the internal market. A proposal for a change in the framework directive was developed (European Union, 1994a), as a consequence of which national regulations for foods for diabetics would have to be deleted. A final decision on the matter has not yet been made.

7.5.1.5 Reduced energy claims. The contribution of polyols to the energy intake of an individual depends not only upon absorption in the small intestine but also, in part, on the extent of colonic absorption of fermentation end products and this is a matter of considerable debate. As an example, some studies have concluded that the energy yield of maltitol is much lower than the 16.7 kJ g^{-1} of sucrose (Oku *et al.*, 1991; Tamura *et al.*, 1991) but others have suggested that there is insignificant energy saving in comparison to sucrose (Beaugerie *et al.*, 1990; Kruger *et al.*, 1992; Zunft *et al.*, 1983). The estimated energy value of maltitol varies from 1.91 kcal g^{-1} (Tsuji *et al.*, 1990) to 3.5 kcal g^{-1} (Beaugerie *et al.*, 1990). However, it is worthy of note that maltitol may also competitively inhibit the hydrolysis of maltose (Nilsson and Jagerstad, 1987; Yoshizawa *et al.*, 1975).

Depending upon national laws worldwide, the energy reduction compared to a normal product has to be between 25% and 50% before a 'reduced energy' claim can be made. Energy values for polyols are laid down in various countries. The EU has laid down a unitary value for all polyols of 2.4 kcal g^{-1} (European Union, 1990) and in the report of the life science research office of the Federation of the American Science and Experimental Biology (FASEB, 1994) it was concluded that fermentation of polyols reaching the colon yields a net energy of 2.0 kcal g^{-1} specifically in the form of short-chain fatty acids.

However, other countries have considered that scientifically based determinations of energy values (Nutrition Council of the Netherlands, 1987; FASEB, 1994) indicate that values for individual polyols are too different from one another to allow agreement on a single value for all polyols. Countries such as the USA and Canada have proposed individual energy values for individual polyols and the FDA has declared that it will not object to the use of the following energy values (kcal g^{-1}) for individual polyols or polyol mixtures:

- hydrogenated starch hydrolysate = 3.0
- sorbitol = 2.6
- xylitol = 2.4
- mannitol, lactitol and isomalt = 2.0

  (sucrose = 4.0)

None of the values above exceeds 3 kcal g^{-1} which is 75% of the energy value of sucrose, thus making it possible for a 'reduced energy' claim to be made for all the polyols shown. In contrast to these observations, Beaugerie *et al.* (1990) investigated the digestion and absorption of maltitol, sorbitol and HGS (Lycasin®) and claimed that the mean estimated energy values in kcal g^{-1} for sorbitol, maltitol and Lycasin® 80/55 were 3.58 ± 0.08, 3.50 ± 0.07 and between 3.11 and 3.54 ± 0.08, respectively suggesting that little calorie saving can be expected from the chronic consumption of these sugar alcohols. Table 7.15 shows the energy values of four polyols determined in a separate study by Beaugerie *et al.* (1991).

7.5.1.6 Low cariogenicity and sugar-free claims. The low cariogenicity of polyols is widely accepted from a scientific point of view (Imfeld, 1994), but from a legal standpoint it is sometimes difficult to make a similar claim for a final product. Several countries regard claims such as non-cariogenic as being forbidden or in need of special permission but allow statements such as 'kind to your teeth'. Table 7.16 shows some of the labels currently carried by sugar-free chewing gums that use xylitol as the sweetening agent.

The terms 'sugar-free' and 'no-sugar-added' are commonly used on products in which sugars are substituted by polyols. The claim is permitted

Table 7.15 Percentage of small intestine malabsorption and energy value of 4 polyols ingested by six fasting subjects (after Beaugerie *et al.*, 1991)

	Lactitol	Sorbitol	Maltitol	Isomalt
Percentage of a 20 g dose of polyol not absorbed in the small intestine				
Mean	84*	66	44**	40**
SEM	14	11	7	7
Energy value				
Mean	2.3	2.7	3.1**	2.8
SEM	0.3	0.2	0.1	0.1

* Different from **; $P < 0.05$.

Table 7.16 Examples of labels currently carried by sugar-free chewing gums using xylitol as the sweetening agent

Company	Brand	Country	Claim
Chocolat Frey	Candida	Switzerland	'Neutralises the acids that attack the enamel of teeth'
Dandy	Dirol	Russia	'Xylitol'
Dandy	Stimorol	Denmark/ Scandinavia	'Sugar-free with Xylitol'
Dandy/Fertin	V6	Scandinavia	'Xylitol and Mouth Hygiene'*
Hitschler	Jetties	Germany	'Sugarless'
Husler	Plax	Switzerland	'Neutralises the acid that can cause plaque and with Xylitol'
Leaf	Xylifresh 100	UK/Belgium/ Netherlands	'Xylifresh 100 helps neutralise plaque acid that can cause cavities and also helps reduce build up of harmful plaque and 100% Xylitol'
Leaf	Xylitol Jenkki	Scandinavia	'Dental Association Endorses Xylitol'
Leaf	Xylifresh	US	'Fights plaque and expressly for not promoting tooth decay'
Leaf	Ben Bits Clinic	Denmark	'Sugar-free with Xylitol'
Lotte	Sugarfree	Korea	'Xylitol'
Malaco	Toy	Sweden	'Xylitol'
Perfetti	Vivadent	Italy	'Xylitol – New Formula'
Stafford Miller	Endekay	UK	'Helps Prevent Tooth Decay'
Warner Lambert	Trident	Canada	'Fights Cavities'
Warner Lambert	Dentyne	Canada	'Fresh new taste'
Wrigley	Extra	Germany	'Cosmetic with Xylitol'
Wrigley	Excel	Canada	'Sugar-free and far longer lasting'
Wrigley	Extra	UK	'Sugar-free'

* Product licence.

in many countries but the conditions for use are governed by many factors. Within the EU there is no final directive on claims but the latest draft from the Consumer Policy Service (European Union, 1994b) does not lay down specific criteria which differ from those of previous drafts. Basically, all claims are to be permitted which are objectively true, not misleading for the consumer, do not discredit other foodstuffs and do not attribute to a food the property of preventing, treating or curing a human disease. Article 8 of the draft forbids more detailed national regulations, a topic which is highly controversial.

7.5.2 *Sugar-free confectionery*

7.5.2.1 General considerations. Because polyols are functional food ingredients, their use is of interest in the design of both reduced calorie foods as well as sugar reduced or sugar-free foods which are kind to teeth. Such non-sugar designer foods are particularly important in confectionery and many examples are now in the marketplace, not just in dietetic outlets, but increasingly in main stream confectionery for the mass market. Sugar confectionery and chocolate countlines are eaten throughout the day and often as snacks as partial replacement for traditional meals. Confectionery is convenient, ready to consume without preparation, easily available, tastes good and is highly gratifying. Although there is clearly no intention or wish to replace all sugar confectionery with reduced or sugar-free alternatives, confectionery manufacturers are well aware of the market opportunities for certain product categories for non-sugar confectionery especially reduced calorie, reduced sugar, tooth-friendly products targeted at health-conscious individuals. There is no doubt that sugar and sugar-free products will increasingly exist side by side, but sugar-containing products are always likely to dominate; after all sugar is a noble nutritious ingredient!

Unfortunately, the first generation of non-sugar confectionery was positioned for diabetics and often tasted awful! Furthermore, most products were based on sorbitol which is renowned for its low threshold for gastrointestinal discomfort and laxation; this latter issue was usually managed by diabetics by limiting the amount of the confectionery which they ate. Nowadays, new generation polyols such as isomalt, maltitol, lactitol and xylitol are widely available and permitted in confectionery. Their inherent physical and chemical properties are not quite the same as the conventional sugars such as sucrose, but food technologists and food process engineers have learnt how to use them and know how to process them in such a manner as to be able to manufacture excellent-tasting products which are indistinguishable and sometimes superior to the sucrose-containing product which is to be replaced.

7.5.2.2 Price constraints. The single most important factor holding back the growth of the non-sugar confectionery market is the high price of new generation polyols. Mainstream consumers are attracted by the positive attributes of non-sugar confectionery but they are not prepared to pay any price for them. The sticking factor within the EU is that new generation disaccharide alcohols cost three to four times the price of sugar and specialist polyols such as xylitol, although used in smaller quantities in a given recipe type, cost over six times the price of sugar. A 25% price premium in mainstream confectionery is generally thought of as being at the maximum that a consumer will pay for an alternative product that is considered to be comparable to the standard product (though there are exceptions). For most confectionery product categories, this purchase price constraint cannot currently be overcome. However, there is a trend towards a reduction in polyol prices and evidence of a growing number of health-conscious consumers. With this in mind, it can be predicted that the non-sugar confectionery market will continue to grow.

7.5.2.3 Products in the marketplace. The most important categories of use of polyols are for chewing gums and candies, but polyol-containing chocolate is emerging as another product in the marketplace. With chewing gum the portion size is typically small, less than 10 g per sale item (disregarding multipacks) and clearly the polyol ingredient is not the major item contributing to the manufacturing cost and the retail price. Sugar-free, tooth-friendly and light (i.e. reduced calorie) claims are very appealing and there is a trend for sugar-free chewing gum to be marketed as a cosmetic item contributing towards dental hygiene and breath freshness. No wonder it is so successful! Data on the chewing gum market (Nielsen, 1993) show that in many countries sugar-free gums have a higher market share than the equivalent sucrose-based gums. In Denmark, Norway and Finland sugar-free gums represent more than 90% of the market and in The Netherlands, Switzerland, Sweden, Spain, Greece, United Kingdom, Belgium, Austria and Germany sugar-free gums account for between 50% and 85% of the market share (Nielsen, 1993). A selection of such products is shown in Figure 7.13.

In the non-gum confectionery market there are also many examples of sugar-free confectionery. The older generation products tended to be large

Figure 7.13 Selection of sugar-free chewing gums.

portion sticks and bags based on sorbitol, although these have largely been superseded by smaller portion sizes. Virtually every type of confectionery product is available in a whole range of flavours and includes hard-boiled sweets, toffees, lozenges and jellies, a selection of which is shown in Figure 7.14. There is also a trend towards increased popularity of sugar-free mint sticks, breath fresheners, cough sweets and nasal decongestants in which the polyol xylitol is used to take advantage of its refreshing effect (due to its negative heat of solution); sorbitol has a similar but less powerful effect. In Switzerland, a country with a highly developed mature confectionery market and with proportionally the largest sugar-free confectionery market, sugar confectionery sales remained static between 1989 and 1993 whilst there was a continuous increase in the sugar-free share from 34.8% to 46.1% (IHA, 1992). In Italy this non-gum sugar-free confectionery market in 1994 accounted for over 20% of market share (Nielsen, 1995). In fact the sugar-free market segment in most other countries, although much smaller, is well established and growing.

Chocolate, in which the sucrose ingredient has been replaced by polyol, has been available for many years although principally positioned for diabetics. In recent years, however, there have been attempts to present sugar-reduced chocolate (with no added sugar) to a wider cross section of consumers. Although cocoa contains very small amounts of sugars, dark

Figure 7.14 Examples of different types of sugar-free confectionery products.

chocolate is considered to be sugar-free; in contrast, however, milk chocolate cannot be classified as such because of the lactose content derived from milk (Rugg-Gunn, 1994). In France there are a number of these products positioned alongside mainstream confectionery and the Poulain product shown in Figure 7.15 is the market leader. In the UK, the supermarket chain Sainsburys have launched reduced-sugar milk chocolate. However, the market segment is extremely small and does not yet account for even 1% of the total UK chocolate market. An interesting development, although still confined to Migros of Switzerland and the dietetic shelves of Boots in the UK, has been the launch of a 'kind to teeth' chocolate bar. The Boots product, positioned for children, is consequently of small portion size (i.e. 30 g) and contains isomalt which is one of the better tolerated polyols (Koutsou *et al.*, 1975).

Acknowledgements

Practical aspects of several of the studies reported here were carried out by research assistants Adam Lee and George Koutsou and their contribution is gratefully acknowledged, as is the original statistical analysis of data attributable to Dr. Phillip Scarfe. These individuals are members of either

Figure 7.15 Three reduced-chocolate products.

The Research and Graduate College or The Science Research Institute of the University of Salford. The authors are grateful to Anke Sentko for useful discussions on the legislative aspects of polyols.

References

Abraham, R.R., Davis, M., Yudkin, J. and Williams, R. (1981) Controlled clinical trial of a new non-calorigenic sweetening agent. *J. Human Nutrition* **35**, 165–172.

Agricultural Research Council/Medical Research Council (1974) *Food and Nutrition Research*, HMSO, London.

Bayless, T.M., Paige, D.M. and Ferry, G.D. (1971) *Gastroenterology* **60**, 605–608.

Bär, A. (1994) "Toothfriendly" – Achievements after 10 years and future prospects. In *Sugarless – Towards the year 2000* (ed. A.J. Rugg-Gunn), Royal Society of Chemistry, Cambridge.

Beaugerie, L. (1987) Contribution à l'étude du transport intestinal du sorbitol chez l'homme. *Thesis*. Hospital St. Lazarre (Paris).

Beaugerie, L., Flourié, B., Franchisseur, C., Pellier, P., Dupas, H. and Rambaud, J-C. (1989) Absorption intestinal et tolerance clinique au sorbitol, mannitol, lactitol et isomalt. *Gastroenterologie Clinique et Biologie* **13(2)**, A102.

Beaugerie, L., Flourié, B., Marteau, P., Pellier, P., Franchisseur, C. and Rambaud, J-C. (1990) Digestion and absorption in the human intestine of three sugar alcohols. *Gastroenterology* **99**, 717–723.

Beaugerie, L., Flourié, B., Pellier, P., Achour, L., Franchisseur, C and Rambaud J-C. (1991) Tolérance clinique, absorption intestinale et valeur énergétique de quatre polyols pris a jeun. *Gastroenterologie Clinique et Biologie* **15**, 929–932.

Billaux, M.S., Flourié, B., Jacquemin, C. and Messing, B. (1991) Sugar alcohols. In *Handbook of Sweeteners* (eds. S. Marie and J.R. Piggott), Blackie, Glasgow, pp 72–103.

BLS (1990) *Bundeslebensmittleschlüssel*, German food data bank, Berlin.

Bond, J.H. and Levitt, M.D. (1972) Use of pulmonary hydrogen measurements to quantitate carbohydrate absorption. *J. Clin. Invest.* **51**, 1219–1225.

Codex Alimentarius Commission (1979) *Procedural Manual*, 7th edn, Food and Agriculture Organization, Rome, p. 31.

Commission of the European Communities (1987) *Report of the Scientific Committee for Food on Sweeteners*, CS/EDUL/56-final, 11 Dec.

Crapo, P.A., Reaven, G. and Olefsky, J. (1976) Plasma glucose and insulin responses to orally administered simple and complex carbohydrate. *Diabetes* **25**, 741–747.

Dahlqvist, A. and Telenius, U. (1965) The utilisation of a presumably low-calorigenic carbohydrate derivative. *Acta physiologica Scand.* **63**, 156–163.

Davenport, H.W. (1978) *A Digest of Digestion*, Year Book Medical Publishers, Chicago.

Department of Health (1989) *Dietary Sugars and Human Health* (Reports on health and social subjects; 37), HMSO, London.

Department of Health (1991) *Dietary Reference Values for Food Energy and Nutrients for the United Kingdom* (Reports on health and social subjects; 41), HMSO, London.

Department of Health (1994) *Nutritional Aspects of Cardiovascular Disease* (Reports on health and social subjects; 46), HMSO, London.

Dwivedi, B.K. (1978) Food products for special dietary needs. In *Low Calorie and Special Dietary Foods* (ed. B.K. Dwivedi), CRC Press, West Palm Beach, Florida, pp. 1–22.

Elliott, J.G. (1994) Functional candies: scientific, marketing and regulatory efforts. *The Manufacturing Confectioner* Dec. 1994, 39–46.

Ertel, N.M. Algun, S., Kemp, F.W. and Mittler, J.C. (1983) The metabolic fate of exogenous sorbitol in the rat. *J. Nutrition* **113**, 566–573.

European Association for the Study of Diabetes (1988) Diabetes and Nutrition Study Group. Nutritional recommendations for individuals with diabetes mellitus. *Diabetes Nutrition Metab.* **1**, 145–149.

European Union (1989a) Council directive of 21 December 1988 on the approximation of the laws of the Member States concerning food additives authorized for use in foodstuffs intended for human consumption, Article 1, *Official Journal of the European Communities* No L. 40/27.

European Union (1989b) Directive 89/389/EEC – Food for Particular Nutritional Uses, *Amblatt der Europäischen Union* Nr. L 186/27.

European Union (1990) Council Directive of 24 September 1990 on nutrition labelling for foodstuffs (90/496/EEC), *Official Journal of the European Communities* No L 276/40.

European Union (1994a) Proposal for a European Parliament and Council Directive amending Council Directive 89/398/EEC on the approximation of the laws of the Member States relating to foodstuffs intended for particular nutritional uses (94/C/108/07), *Official Journal of the European Communities* No C 108/17.

European Union (1994b) Draft Commission Directive amending Directive 94/54/EC concerning the compulsory indication on the labelling of certain foodstuffs of particulars other than those provided for in Directive 79/112/EEC. III/5518/94–Rev.J–EN.

Felber, J.P., Tappy, L., Vouillamoz, D., Randin, J.P. and Jequier, E. (1987) Comparative study of maltitol and sucrose by means of continuous indirect calorimetry. *J. Parenteral Enteral Nutrition* **11** (3), 250–254.

Flourié, B., Leblond, A., Florént, C., Rautureau, M., Bisalli, A. and Rambaud, J-C. (1988a) Starch malabsorption and breath gas excretion in healthy humans consuming low- and high-starch diets. *Gastroenterology* **95**, 356–363.

Flourié, B., Florent, F., Etanchaud, F., Evard, D., Franchisseur, C. and Rambaud, J–C. (1988b) Starch absorbance by healthy man evaluated by lactulose hydrogen breath test. *Am. J. Clin. Nutrition* **47**, 61–66.

Food and Drug Administration (1994) Frozen desserts. *Federal Register* **59**, (177), 47072 ff.

Fondu, M. and Gerard, A. (1991) The legal status of novel foods. *Alimentalex. Internat. Food Law Rev.* **5**, 3–40.

Fritz, M., Siebert, G. and Kaspar, H. (1985) Dose dependence of breath hydrogen and methane in healthy volunteers after ingestion of a commercial disaccharide mixture, Palatinit®. *Br. J. Nutrition* **54**, 389–400.

Goda, T., Takase, S and Hosoya, N. (1988) Hydrolysis of alpha-D-glucopyranosyl-1,6-sorbitol and alpha-D-glucopyranosyl-1,6-mannitol by rat intestinal disaccharidases. *J. Nutrition. Sci. Vitaminology* **34**, 131–140.

Goldberg, I. (1994) *Functional foods – Designer foods, Pharmafoods, Nutraceuticals.* Chapman & Hall, New York.

Goossens, J. and Roper, H. (1993) Erythritol, a new sweetener. In *Carbohydrates as Organic Materials II* (ed. Descotes), VCH, Weinheim and New York, pp 27–38.

Goovaerts, L. and Ravelli, G.P. (1993) Lactitol monohydrate for the treatment of chronic constipation: A multicentre study of the efficacy and tolerability of an individually adjusted daily dose. *Acta Therapeutica* **19**, 61–71.

Grenby, T.H. and Phillips, A. (1989) Dental and metabolic effects of lactitol in the diet of laboratory rats. *Br. J. Nutrition* **61**, 17.

Grimble, G.K., Patil, D.H. and Silk, D.B.A. (1988) Assimilation of lactitol, and 'unabsorbed' disaccharide in the normal human colon. *Gut* **29**, 1666–1671.

Grupp, U. and Siebert, G. (1978) Metabolism of hydrogenated palatinose, an equimolar mixture of alpha-D-glucopyranosido-1,6-sorbitol and alpha-D-glucopyranosido-1,6-mannitol. *Res. Exp. Med. (Ber.)* **173**, 261–278.

Gryboski, J.D. (1966) Diarrhoea from diatetic candies. *New Engl. J. Med.* **29**, 718.

Hammer, H.F. (1993) Colonic hydrogen absorption: quantification of its effect on hydrogen accumulation caused by bacterial fermentation of carbohydrates. *Gut* **34**, 818–822.

Hammer, H.F., Santa-Anna, C.A., Schiller, L.R. and Fordtran, J.S. (1989) Studies of osmotic diarrhoea induced in normal subjects by ingestion of polyethylene glycol and lactulose. *J. Clin. Invest.* **84**, 1056–1062.

Heitland, W. and Mauersberger, H. (1988) A clinical investigation comparing the laxative effect of lactitol to that of lactulose in a randomised, open, parallel study. *Schweiz Rundschau Med.* **77**, 493–495.

Hiele, M., Ghoos, Y., Rutgeerts, P. and Vantrappen, G. (1993) Metabolism of erythritol in humans: comparison with glucose and lactitol. *Br. J. Nutrition* **69**, 169–176.

Hyams, J.S. (1983) Sorbitol intolerance: an unappreciated cause of functional gastrointestinal complaints. *Gastronterology* **84**, 30–33.

IHA (1992) *Candy Market Switzerland*, GFM (Institut für Marktanalysen AG mit GFM Forschungs-institut der Schweizerischen Gesellschaft für Marketing), Switzerland.

Imfeld, T.M. (1994) Clinical caries studies with polyalcohols. A literature review. *Schweiz Monatsschr. Zahnmed.* **104** (8), 941–945.

Johnson, J.D. (1981) The regional and ethnic distribution of lactose malabsorption. In *Lactose Digestion* (eds. D.M. Paige and T.M. Bayless), Johns Hopkins University Press. Baltimore, London, pp. 11–22.

Johnson, L.R. (1991) *Gastrointestinal Physiology*, Mosby Year Book, 4th ed., St. Louis.

Kamoi, M. (1975) Study on metabolism. Part 1: Fundamental experiment. *J. Jpn Diab. Soc.* **18**, 243–249.

Kearsley, M.W. and Birch, G.G. (1978) Blood glucose profiles in man after ingestion of hydrogenated glucose syrups. *IRCS Med. Sci.* **6**, 82.

Kearsley, M.W., Birch, G.C. and Lian-Loh, R.H.D. (1982) The metabolic fate of hydrogenated glucose syrups. *Stärke* **34**, 279–83.

Koizumi, N., Fuji, M., Ninomiya, R., Inoue, Y., Kagawa, T. and Tsukamoto, T. (1983) Studies on transitory laxative effects of sorbitol and maltitol. 1. Estimation of 50% effective dose and maximum non-effective dose. *Chemosphere* **12(1)**, 45–53.

Koutsou, G.A., Lee, J.A., Zumbé, A. and Storey, D.M. (1995) Dose-related gastrointestinal response to the ingestion of either isomalt, lactitol or maltitol in chocolate. *European J. Clin. Nutrition* in press.

Krüger, D., Grossklaus, R., Klingbeil, L., Ziese, T and Kock-Gensecke, S. (1991) Caloric availability of Palatinit® (isomalt) in the small intestine of rats: implications of dose dependency on the energy value. *Nutrition Res. Rev.* **11**, 669–678.

Krüger, D., Grossklaus, R., Herold, M., Lorenz, S. and Klingbeil, L. (1992) Gastrointestinal transit and digestibility of maltitol, sucrose and sorbitol in rats: a multicompartmental model and recovery study. *Experientia* **48**, 733–740.

Labell, F. (1993) Europe/Japan open doors to medical food products. *Food Production* **54** (4) 60.

Lauwers, A.M., Daumerie, C. and Henquin, J.C. (1985) Intestinal absorption of sorbitol and effects of its acute administration on glucose homeostasis in normal rats. *Br. J. Nutrition* **53**, 53.

Le Bot, Y. (1993) Update on bulk sweeteners polyols. In *Food Ingredients Europe*, Conference Proceedings, 25–27th November, Expoconsult Publishers, Dusseldorf, pp. 291–296.

Lee, A., Zumbé, A. and Storey, D.M. (1994) Breath hydrogen after ingestion of the bulk sweeteners sorbitol, isomalt and sucrose in chocolate. *Br. J. Nutrition* **71**, 731–737.

Lian-Loh, R., Birch, G.G. and Coates, M.E. (1982) The metabolism of maltitol in the rat. *Br. J. Nutrition* **48**, 477–481.

Life Science Research Office (1994) *The Evaluation of the Energy of Certain Sugar Alcohols used as Food Ingredients*, Federation of American Societies for Experimental Biology (FASEB). Life Science Research Office, Bethesda, Maryland (USA).

Livesey, G. (1992) The energy values of dietary fibre and sugar alcohols for man. *Nutrition Res. Rev.* **5**, 61–84.

Livesey, G., Johnson, L.T., Gee, J.M., Smith, T., Lee, W.E., Hillan, K.A., Meyer, J. and Turner, S.C. (1993) 'Determination' of sugar alcohol and Polydextrose® absorption in humans by the breath hydrogen (H_2) technique: the stoichiometry of hydrogen production and the interaction between carbohydrates assessed *in vivo* and *in vitro*. *European J. Clin. Nutrition* **47**, 419–430.

MAFF (1990) *Household Food Consumption and Expenditure 1989*, HMSO, London.

McNeil, N.I. (1984) The contribution of the large intestine to energy supplies in man. *Am. J. Clin. Nutrition* **39**, 338–342.

McNeil, N.I., Cummings, J.H. and James, W.P.T. (1978) Short-chain fatty acid absorption by the human large intestine. *Gut* **19** 819–822.

Menzies, I.S. (1983) Medical importance of sugars in the alimentary tract. In *Developments in Sweeteners*, Vol. 2 (ed. T.H. Grenby, K.J. Parker and M.G. Lindley) Applied Science, London and New York, pp. 89–117.

Metzger, J., Chollet, C., Wermeille, M., Biollaz, J., Llull, J.B. and Lauterburg, B.H. (1988) Lactitol: Gastrointestinal absorption and effect on blood lactate in healthy volunteers and patients with cirrhosis. *European J. Clin. Pharmacol.* **35**, 97–99.

Miller, T.L. and Wolin, M.J. (1979) Fermentations by saccharolytic intestinal bacteria. *Am. J. Clin. Nutrition* **32**, 164–174.

Moskowitz, A.H. (1980) Maltitol and Hydrogenated Starch Hydrolysate. In *Intense Sweeteners*, chapter 14, pp. 259–283.

Nasrallah, S.M. and Iber, F.L. (1969) Mannitol absorption and metabolism in man. *Am. J. Med. Sci.* **258**, 80–88.

Nielsen (1993) *Chewing Gum Market Shares*, Frankfurt/M., Germany.

Nielsen (1995). Document. St-85-SPE-SPEZ, AL. 06.02.St-MA-AKJ-88. Nielsen, Italy.

Nilsson, U and Jagerstad, M. (1987) Hydrolysis of lactitol, maltitol and Palatinit® by intestinal biopsies. *Br. J. Nutrition* **58**, 199–206.

Nutrition Council of the Netherlands (1987) *The Energy Value of Sugar Alcohols*, Recommendations of the Committee on Poly-alcohols, The Hague.

Oku, T. and Noda, J. (1990) Influence of chronic ingestion of a newly developed sweetener, erythritol on growth and gastrointestinal function of the rat. *Nutritional Res.* **10**, 987–996.

Oku, T., Inoue, Y and Hosoya, N. (1971) Absorption and excretion of maltitol-U-^{14}C in the rat. *J. Jpn Soc. Food Nutrition* **24**, 399–404.

Oku, T., Akiba, M., Lee, M.H., Moon, S.J. and Hosoya, N. (1991) Metabolic fate of ingested [U^{14}C]-maltitol in man. *J. Nutrition. Sci. Vitaminology* **37**, 529–544.

Oliger, P.M. (1994) New options for sucrose-free chocolate. *The Manufacturing Confectioner*, May, 77–84.

Paige, D.M., Bayless, T.M. and Davis, L.R. (1975) Lactose intolerance and lactose hydrolyzed milk. In *Physiological Effect of Food Carbohydrates* (ed. A. Jeans and J. Hodge), ALS Symposium Series 15, American Chemical Society, Washington D.C., pp. 191–206.

Patil, D.H., Grimble, G.K. and Silk, D.B.A. (1987) Lactitol, a new hydrogenated lactose derivative: intestinal absorption and laxative threshold in normal human subjects. *Br. J. Nutrition* **57**, 195–100.

Pometta, D. and Trabichet, C. (1985) Rapport concernant l'utilisation du Palatinit® chez les diabetiques de type II traités par la régime seul Hôpital Cantonal Universitaire de Génève. *Akt. Ernähr* **10**, 174–177.

Rambaud, J-C. and Flourié, B. (1994) Mechanism of carbohydrate-induced diarrhoea. In *Short Chain Fatty Acids* (eds. H.J. Binder, J.H. Cummings and K. Soergel) pp. 232–239.

Rappaille, A., Gonze, M. and Van der Schueron, F. (1994) The ideal sweetener for sugar-free chocolate. *Confectionery Production* May, 331–336.

Rennhard, H.H. and Bianchine, J.R. (1976) Metabolic and caloric utilization of orally administered maltitol C^{14} in the rat, dog and man. *J. Agric. Chem.* **24**, 287–290.

Rosiers, C., Verwaerde, F., Dupas, H. and Bouquelet, S. (1985) New approach to the metabolism of starch hydrolysate: hydrolysis by the maltase glucoamylase complex of the rat intestinal mucosa. *Annals Nutrition Metabolism* **29**, 76–82.

Rugg-Gunn, A. (ed) (1994) *Sugarless: Towards the year 2000*, Royal Society of Chemistry, Cambridge.

Rumessen, J.J., Hamberg, O. and Gudman-Hoyer, E. (1990) Interval sampling of end-expiratory hydrogen (H_2) concentrations to quantify carbohydrate malabsorption by means of a lactulose standard. *Gut* **31**, 37–42.

Sanford, P.A. (1992) *Digestive System Physiology*, Edward Arnold, London.

Saunders, D.R. and Wiggins, H.S. (1981) Conservation of mannitol, lactulose and raffinose by the human colon. *Am. J. Physiol.* **241**, G397–G402.

Shindou, T., Sasaki, Y., Miki, H., Eguchi, F., Hagiwara, K. and Ichikawa, T. (1989) Identification of erythritol by HPLC and GC-MS and quantitative measurement in pulps of various fruits. *J. Agric. Food Chem.* **37**, 1474–1476.

Sicard, P.J. and Le Bot, Y. (1994) Manufacturing opportunities with non-sugar sweeteners, In *Sugarless – Towards the Year 2000* (ed. A.J. Rugg-Gunn), Cambridge, The Royal Society of Chemistry, Cambridge, pp. 112–135.

Sicard, P.J. and Leroy, P. (1983) Mannitol, sorbitol and lycasin: properties and food applications. In *Developments in Sweeteners 2*, (ed. T.H. Grenby) Applied Science.

Souci, S.W., Fachmann, W. and Kraut, H. (1989) *Food Composition and Nutrition Tables*, 4th revised and completed edition compiled by H. Scherz and F. Senser, Wissenschaftliche Verlagsgesellschaft mbH, Stuttgart.

Spengler, M., Somogyi, J.C., Pletcher, E. and Boehme, K. (1987) Tolerability, acceptance and energetic conversion of isomalt (Palatinit®) in comparison with sucrose. *Akt. Enrähr.* **12**, 210–214.

SRU (1992) *Consumer Food Tolerance Survey Switzerland*, London.

Stein, A. (1992) Aspects de definitions legales dans la legislation alimentaire. Alimentalex. *Internat. Food Law Rev.* **8**, 11–16.

Storey, D.M., Zumbé, A. and Lees, A. (1994) Polyols, breath hydrogen and fermentation. *Brit. J. Nutr.* **72**, 947–952.

Tamura, Y., Furuse, M., Matsuda., Shimizu, T. and Okumura, J. (1991) Energy utilisation of sorbose in comparison with maltitol in growing rats. *J. Agric. Food Chem.* **39**, 732–735.

Thiébaud, D., Jacot, E., Schmitz, H., Spengler, M. and Felber, J.P. (1984) Comparative study of isomalt and sucrose by means of continuous indirect calorimetry. *Metabolism* **33** (9), 808–813.

Tsuji, K., Osada, Y., Simada, N., Nishimura, R., Kobayashi, S., Ichikawa, T. and Hoyosa, N. (1990) Energy production of sorbitol and maltitol in healthy men and rats. In *Caloric Evaluation of Carbohydrates* (ed. N. Hoyosa), The Japan Association of Dietetic and Enriched Foods, Tokyo, pp. 77–90.

Ugolev, A.N. (1974) *Gut*, **13**, 735–747.

Van Es, A.J.H., De Groot, L. and Vogt, J.E. (1986) Energy balances of eight volunteers fed on diets supplemented with either lactitol or saccharose. *Br. J. Nutrition* **56**, 545–554.

Van Weerden, E.J. and Huisman, J. (1993a) The digestion process of the sugar alcohol isolmalt in the intestinal tract of the pig. 1 – Studies with administration of isomalt in the feed. *Br. J. Nutrition* **69**, 455–466.

Van Weerden, E.J. and Huisman, J. (1993b) The digestion process of the sugar alcohol isomalt in the intestinal tract of the pig. 2 – Studies with administration of isomalt as a sweet. *Br. J. Nutrition* **69**, 467–474.

Verwaerde, F. and Dupas, H. (1982) Study of the *in-vitro* digestion of Lycasin® 80/55 in the rat. Unpublished report from Roquette Frères, Lestrem, France. Submitted to WHO by Roquette Frères, 1982.

Vincent, M. (1989) *Les Polyols: de l'additif à l'ingrédient*, Seminaire "Novel Foods – Innovations et sécurité Alimentaires", Centre de Perfectionnement des Cadres des Industries Agricoles et Alimentaires (C.P.C.I.A.), Paris.

Wang, Y.M. and Van Eys, J. (1981) Nutritional significance of fructose and sugar alcohols. *Ann. Rev. Nutrition* **1**, 437–475.

WHO (1983) Toxicological evalution of certain food additives and food contaminants. 27th Report of the Joint FAO/WHO Expert Committee on Food Additives. *WHO Technical Report Series*, No. 696. Geneva: WHO.

WHO (1990) *Evaluation of Certain Food Additives and Contaminants*, 35th Report of the Joint FAO/WHO Expert Committee on Food Additives, World Health Organization, Geneva.

Willibald-Ettle, I. (1994) Technical Application Servies, Palatinit® Süßungsmittel GmbH, Mannheim, FRG, personal communication.

Wursch, P. and Ananthamaran, G. (1989) Aspects of the energy value assessment of the polyols. In *Progress in Sweeteners* (ed. T.H. Grenby) Elsevier Applied Science, London and New York, pp. 241–266.

Wursch, P. and Del Vedovo, S. (1981) Inhibition of human digestive enzymes by hydrogenated malto-oligosaccharides. *Internat. J. Vitaminology Nutrition Res.* **51**, 161–165.

Wursch, P., Koellreuter, B. and Schweizer, T.F. (1989) Hydrogen excretion after ingestion of five different sugar alcohols and lactulose. *European J. Clin. Nutrition* **43**, 819–825.

Wursch, P., Koellreuter, B., Getaz, F. and Arnaud, M.J. (1990) Metabolism of maltitol by conventional rats and mice, and germ-free mice, and comparability between maltitol and sorbitol in germ-free mice. *Br. J. Nutrition* **63**, 7–15.

Yoshizawa, S., Moriuchi, S. and Hoyosa, N. (1975) The effects of maltitol on rat intestinal disaccharides. *J. Nutrition Sci. Vitaminology* **21**, 31–39.

Ziesenitz, S.C. and Siebert, G. (1987) The metabolism and utilisation of polyols and other

bulk sweeteners compared with sugar. In *Developments in Sweeteners*, vol 3, (ed. T.H. Grenby), Elsevier Applied Science, Barking, Essex, pp. 109–149.
Zumbé, A. and Brinkworth, R.A. (1992) Comparative studies of gastrointestinal tolerance and acceptability of milk chocolate containing either sucrose, isomalt or sorbitol in healthy consumers and type II diabetics. *Zeit. Ernähr.* **31**, 40–48.
Zumbé, A., Lee, A. and Storey, D.M. (1994) Manufacture and marketing of non-sugar chocolate. In *Sugarless – Towards the Year 2000* (ed. A.J. Rugg-Gunn), The Royal Society of Chemistry, Cambridge, pp. 147–171.
Zunft, H-J., Schulze, J., Gartner, H. and Grutte, F-K. (1983) Digestion of maltitol in man, rat and rabbit. *Annals Nutrition Metabolism* **27** (6), 470–476.

8 Glucose syrups in the fermentation industries

D. WOOD and T. O'ROURKE

8.1 Introduction

This chapter discusses the commercial use of glucose syrups in industrial fermentation processes in the United Kingdom. It is split into two sections, the first covering the use of glucose syrups in the production of food ingredients, processing aids, chemicals and pharmaceuticals. The second part covers the history and use of glucose syrups in the United Kingdom brewing industry.

Glucose syrups are used widely in the fermentation industries in the United Kingdom. Each year over 100 000 tonnes of syrups are sold into areas of the fermentation industry including brewing, pharmaceuticals, chemicals, cider, wine and sherry, novel foods, and gums. This represents a significant percentage of the total UK glucose syrup usage.

8.1.1 Production

Glucose syrups are products of the hydrolysis of starch into mixtures of D-glucose polymers. The most common glucose syrup described here is a product composed almost entirely of dextrose and has a Dextrose Equivalent (DE) of 90 or more. This provides the most readily usable form of fermentable sugar for the processes described herein. It should be noted that such highly converted glucose syrups are often referred to as hydrosylates and the term is used synonymously in the industry.

Glucose syrups of 95 DE are produced by the enzymic hydrolysis of starch, using carbohydrases such as α-amylase and amyloglucosidase to break down the starch source predominantly to dextrose. Other enzymes can be used to produce different carbohydrate spectra according to the needs of the end user, such as the maltose-rich syrups commonly used in the brewing industry. Examples of the range of products available are shown in Table 8.1.

Note that some of the industrial processes discussed below have a high proportion of freely available information on the process involved and the advantages of the types of glucose syrups used. Other industries consider this information proprietary and the degree of depth of each section will reflect this.

Table 8.1 Carbohydrate spectra of typical glucose syrups

Sugar	20 DE	42 DE	63 DE	95 DE	High maltose	Very high maltose
Dextrose	1.5	19	37	92	3	3
Maltose	7.0	14	32	5	55	70
Maltotriose	11.0	11	12	3	15	20
Higher sugars	56.0	56	19	0	27	7

8.1.2 Background

The history of glucose syrups has been discussed in many previous books, but it is worth noting that by 1870 a large industry was already established in Germany and France for the production of glucose syrups (Mulvihill, 1992), the applications of which included wine manufacture and brewing. By the 1920s, progress in dextrose crystallisation had made available anhydrous and monohydrate forms at prices competitive with sucrose. Production of glucose had become a large and stable industry.

More recently major advances in enzyme and process technology have allowed glucose syrup manufacturers to produce glucose syrups which can be tailored to specific customer needs, whether they are for very high dextrose, high maltose or fully soluble low conversion syrups to suit the specific needs of the organism or process involved.

8.2 Production of food ingredients and additives

8.2.1 Gums

Traditionally gums (polysaccharides) are used in the food industry to provide a wide range of functional benefits such as thickening and gelling. Natural gums such as acacia gum or gum tragacanth are extracted from plants. The seasonal fluctuations in price and supply of these have made the advent of industrially produced fermentation products such as xanthan gum and gellan gum attractive alternatives. Often blends of gums are used to achieve the desired properties. For instance, xanthan forms a thermo-reversible cohesive gel system with locust bean gum.

Xanthan gum is a mixed polysaccharide with a remarkable resistance to heat and pH changes. It is produced on a wide scale in the world from the fermentation of glucose syrup. The type of glucose syrup used will depend on the specificity of the strain of *Xanthomonas campestris* used, and the availability and price of local glucose syrups. In the USA 42 DE syrups have traditionally been used as the most cost effective carbon source. It is known that both 63 DE and high fructose syrups can be used, again dependent on the supply economics and gum yields involved. Syrups used need to be low in colour to ensure the final product is low in colour itself, so substrates such as molasses would not normally be suitable.

Xanthan gum is produced by aerobic submerged fermentation (Kelco, 1994). The fermentation medium contains glucose syrup, a nitrogen source and trace elements. It is a batch process where the key parameters such as pH, temperature, and aeration and agitation are closely controlled. On completion of this stage the broth is pasteurised and the gum is recovered by precipitation through the addition of propan-2-ol. The alcohol is then removed and the product is dried, milled and sieved to a specified particle size, tested and packaged.

8.2.2 *Novel foods*

Novel foods (ICI, undated) derived from fermentation have had a rather chequered history. Products such as protein derived from methanol/ammonia fermentations gave a great deal of experience to the design and running of continuous fermenters, but were a mixed success commercially.

More recently the advent of mycoprotein on to the UK market has shown signs of promise. Mycoprotein is a protein and fibre-rich mycelial food produced by the continuous fermentation of a selected strain of *Fusarium graminearum* on a high DE glucose substrate. The process is shown in Figure 8.1. Note that the steps from sterilisation through to RNA reduction are carried out under conditions of sterility.

Following RNA reduction and filtration mycoprotein emerges as a filter cake of about 30% solids, which is subsequently texturised and flavoured to make a range of meat substitutes. The type of glucose syrup needed for the process is a deionised 95 DE syrup. As the process is sensitive to changes in substrate, generally fermentations are restricted to one source

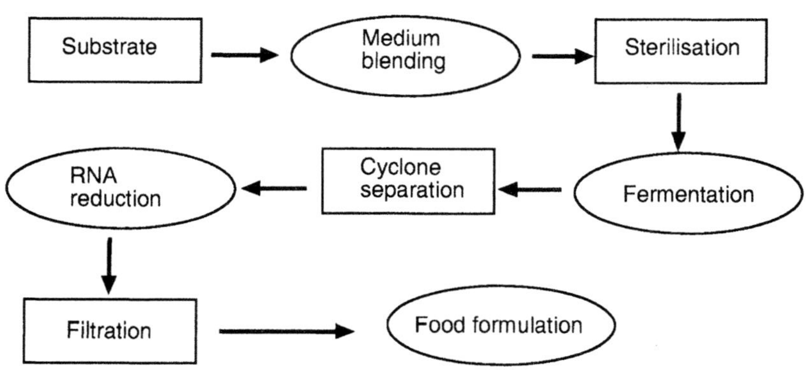

Figure 8.1 Production of mycoprotein.

of supply to maintain high yields. Fermentations can run for 500 to 1000 hours before a morphological change occurs which renders the final product incapable of being texturised. The process then has to be shut down, the plant cleaned and sterilised, and restarted. Deionised glucose syrups are needed for the following reasons:

- The mycoprotein is only filtered to remove water, thus any substrate characteristics such as colour would be likely to transfer over to the finished product.
- The mycoprotein has a very high flavour absorbency, and thus can pick up flavour notes from the substrate.
- High chloride levels in a syrup can give rise to stress corrosion in stainless steel fermenters over a period of time.

8.2.3 Chemicals

Glucose syrups have long been used in fermentations to produce acidulants, such as citric acid (Haarman and Reimer, undated), fumaric acid, propionic acid and lactic acid (Rhodes and Fletcher, 1966). Citric acid is by far the largest in commercial terms and is the product covered in most detail here. Gluconic acid is also produced commercially.

8.2.3.1 Lactic acid. Lactic acid is made by either chemical synthesis or by fermentation of glucose. In the fermentation process glycolysis of the glucose syrup (preferably 95 DE, concentration 5–20% w/v) is carried out by the organism to produce pyruvic acid, which under anaerobic conditions is broken down to produce lactic acid. The organism is a bacterium such as *S. lactis* and is microaerobic. The fermentation takes place at 40–50°C over about 2–3 days and needs no aeration. The pH is continuously controlled to counteract the acidity of the lactic acid. Yields of 90% from the 95 DE syrup used are commonly achieved. Contamination by other organisms is not usually a problem as high temperature and lack of oxygen prevent this. Separation then takes place to produce a material suitable for use in sauces, pickles, drinks, etc. It is also used in detergents as a substitute for polyphosphates, and in other cleaning products.

8.2.3.2 Citric acid. Citric acid is one of the most widely used food acidulants. Initially it was extracted from natural sources, such as lemons. Demand outstripped supply up until the 1920s, when the large scale commercial production of citric acid by *Aspergillus niger* was developed.

Citric acid is an essential ingredient in soft drinks, food, confectionery and pharmaceuticals. It is produced in two basic forms, anhydrous and monohydrate.

Initially the surface process was used, where the *A. niger* is grown on trays containing media of sterilised beet molasses and inorganic salts housed in ventilated rooms. A sterile air stream supplies oxygen and simultaneously cools the fermentation. The mould forms a mycelium layer on the liquid surface of the flat trays, where after around 10 days, the mycelium layer is removed and the citric acid is precipitated as calcium citrate. This is then filtered and washed, regenerated as citric acid, decolorised, concentrated, dried, sieved and packed. A significant proportion of the world's citric production is still made by this method.

After the Second World War submerged fermentation processes were developed and today 80% of world production is made by this type of process which is carried out as an aerobic fermentation in tower fermenters. It is favoured owing to:

- its ability to use a wide range of sugar containing raw materials
- higher yield of citric acid
- improved process control and shorter time
- lower labour and space costs
- better control of sterility

Separation of the citric acid is similar to the surface process. Raw materials for the submerged citric acid production process include sucrose, raw sugar, starch, glucose syrups, beet molasses and maize grits. Choice of raw material generally depends on local economic factors. Exceptions to this occur where certain advantages can be gained through the use of purer raw materials, for example partial substitution of a relatively impure substrate such as molasses by glucose syrup to increase output, reduce recovery and purification costs and lower the effluent loading. The most popular type of glucose syrup is a 95 DE product, where maximum fermentability is needed to ensure maximum yield. Citric acid producers are known to purchase raw starch and convert this on site to a 95 DE product. A fully deionised and filtered syrup is not needed for this fermentation, so a partially filtered glucose syrup is particularly cost effective.

8.2.3.3 Ethanol. Ethanol has been produced from fermentation processes for thousands of years. Most ethanol for use as a chemical feedstock is produced by chemical means, primarily through the catalytic hydration of ethylene. In the light of concern over energy shortages, the production of ethanol from carbohydrates such as glucose syrup has been studied extensively. Under anaerobic fermentation conditions and high glucose concentrations *S. cerevisiae* produces ethanol, once optimum biomass concentrations have been produced under aerobic conditions. Many different sources of carbohydrate can be used, however 95 DE glucose is the most common form of glucose syrup in use. Degree of refining of the carbohydrate substrate is not normally important, as long as it is of a consistent quality.

8.2.3.4 Gluconic acid. Gluconic acid is produced from glucose syrup by microbial oxidation via *Glucose oxidase*. High DE syrups are used to produce the gluconic acid via a submerged fermentation process using *Aspergillus niger* or *Acetobacter suboxydans*. Gluconic acid is used in the manufacture of leather and foods, and also in the processing of certain metals.

8.2.3.5 L-Ascorbic acid. The production of L-ascorbic acid (vitamin C) consists of several chemical steps and one microbial conversion. The steps are shown in Figure 8.2. The oxidation step from D-sorbitol to L-sorbose is carried out by *Acetobacter suboxydans* in a submerged process at 30–35°C with aeration and agitation. Deionised 95 DE glucose syrup is used as the raw material for the process.

8.2.4 Pharmaceutical products

Many pharmaceutical products are derived from fermentations based on glucose syrups, such as penicillin, cephalosporin and griseofulvin. Worldwide in this application the consumption amounts to over 150 000 tonnes of glucose syrup per year. The substrate generally used is a 95 DE syrup, as most of the organisms use only the dextrose fraction of the syrup, so it must have as high DE as practicable. Deionisation is not always necessary as salts and growth factors are added to the fermentation. The colour of the syrup is important where a light coloured end product is required.

Certain intermediate products to be grafted onto traditional antibiotics

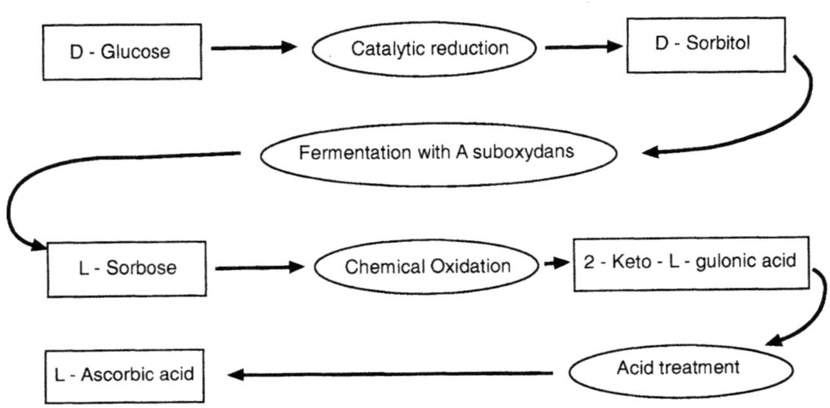

Figure 8.2 Production of vitamin C (L-ascorbic acid).

such as penicillin are fermented from very low DE syrups. This is due to a soil-based organism being used which prefers high molecular weight substrates naturally found in its normal habitat. The type of glucose used is typically a 20 DE product with a high proportion of oligosaccharides.

8.2.4.1 Griseofulvin. Griseofulvin is a non-toxic systemic antifungal antibiotic for the treatment of fungal diseases by oral therapy, such as ringworm. It is produced by submerged fermentation using a strain of *Penecillium griseofulvin*. A 95 DE glucose syrup is used as the carbohydrate source and is fed into the medium on an intermittent basis. Good aeration is necessary, along with careful control of available nitrogen, pH and the concentrations of phosphate and calcium carbonate. Addition of the glucose syrup causes the pH to rise as consumption of the glucose takes place, but this can be controlled by the addition of more carbohydrate.

8.2.4.2 Growth promoters. Growth promoters for the animal feed industry can be produced from a fermentation process using 95 DE glucose syrup as the main carbohydrate source. Once the fermentation is complete, the broth is spray dried, packed and sold. The growth promoters are believed to encourage the growth of acid bacteria in the animal gut and slow the growth of the methane producing bacteria thereby enhancing growth rates of the animal.

8.2.4.3 Enzymes. Enzyme production by fermentation constitutes a huge market throughout the world, with high growth rates. Around half the enzyme production in the West is centred in Denmark. Most enzymes are hydrolytic in action and are used for the depolymerisation of natural products. Many are proteolytic, for use in the dairy, detergent and leather industries. The baking, brewing, distilling, textile and starch industries all use considerable quantities of carbohydrases.

Little information is freely available on the fermentation processes used in the manufacture of enzymes, but it is known that 95 DE syrups, usually deionised, are used as the carbohydrate substrate for many fermentations.

8.3 Use of syrups in brewing

8.3.1 Introduction

The earliest records of brewing date back to Egyptian times, over 4000 years ago. The original beverages were made from cereals, which were probably allowed to germinate in order to produce the enzymes necessary to hydrolyse the starch to fermentable sugar before addition of the yeast to convert the sugars into alcohol.

By the Middle Ages the process of brewing had progressed so that the grain was germinated and dried to produce malt (Figure 8.3), which could then be crushed and mixed with water at a suitable temperature (ca. 65°C), to produce a mash allowing the starch from the grain, usually barley, to be broken down into sugars prior to fermentation. It was found that other carbohydrate sources which were readily available and often cheaper as a

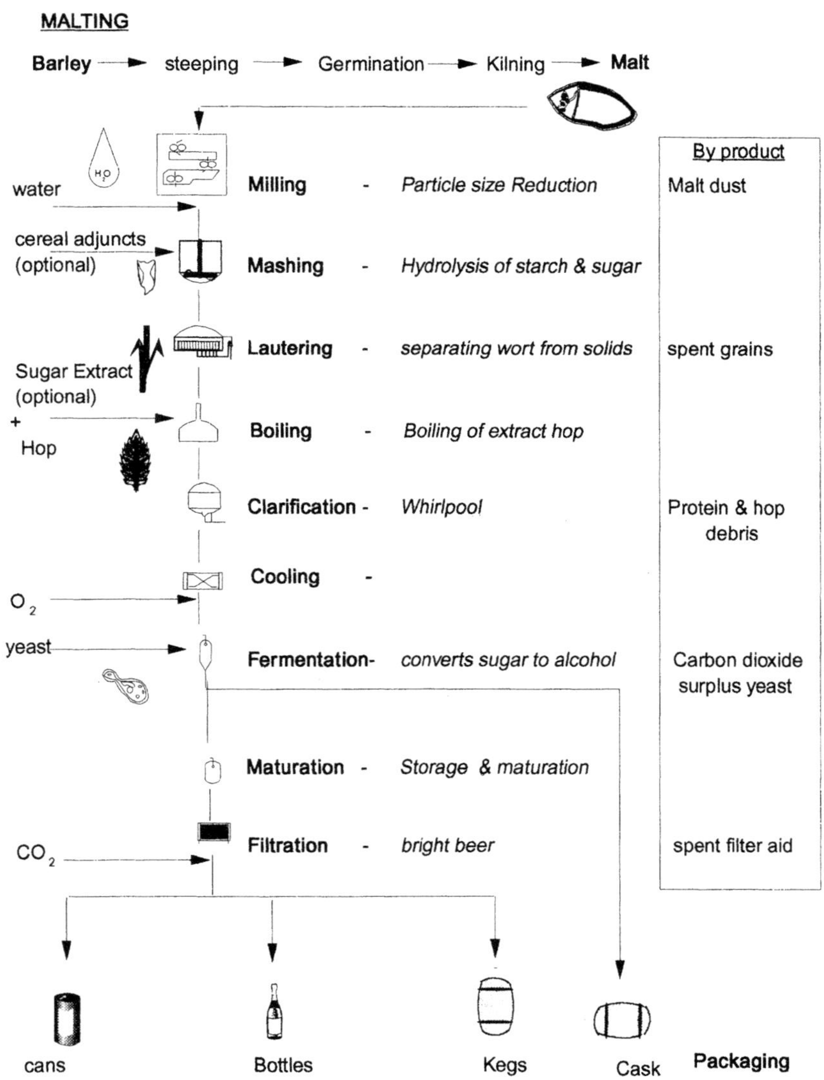

Figure 8.3 Schematic diagram of brewing process.

source of extract (fermentable carbohydrate) could be used to augment or replace some of the malt.

Adjuncts (unmalted carbohydrates) can be generally classified into solid adjuncts, usually other cereals which are a pure source of starch and use the enzymic capacity of the malt to convert the starch to sugars. It is often necessary to cook or boil the starch in a separate vessel to gelatinise it before blending it with the malt mash to complete the saccharification or conversion of the starch to fermentable sugars. Alternatively sugars and syrups can be used which come either directly from sugar sources such as honey, cane or beet, or are produced by a sugar refiner from the external hydrolysis of cereal starches with acid or enzymes.

For many centuries the only available source of sugar was honey, which is still used to produce its own fermented beverage, mead, as well as speciality beers such as Bragot's ale. It was also used to sweeten beers. In the nineteenth century treacle and molasses were used to produce dark rich full flavoured ales, and it is not until the twentieth century that clearer, lighter, more purified sugars and syrups became available.

As well as a source of additional fermentable extract, sugars contribute other properties, which give the essential character to certain styles of beer. For example many milds and sweet stouts contain caramel (burnt sugar) for both flavour and colour. Sugars can be added at the end of primary fermentation to provide sweetness or additional extract to promote a secondary fermentation.

In addition to the amount of adjunct used, there is an interesting difference between the types of adjunct favoured in different parts of the world. Most American and European brewers use solid non-gelatinised adjuncts, whilst in the UK and Australia sugar or syrup adjuncts are preferred. This may be partially explained by the availability of the different raw materials, but it is also a consequence of history and legislation and has had a major influence on the development of brewing in different countries.

8.3.2 History of the use of adjuncts

Adjuncts have not always been permitted in the brewing of beer, especially as brewing has consistently been exploited as a source of revenue for governments. A traditional method of raising taxes both in the UK and Germany was on malt, and hence there was strict legislation prohibiting brewers from using adjuncts to evade duty when taxation was based on the weight of malt used.

The best known legislation is the German 'purity law' or 'Reinheitsgebot', first introduced in Munich in 1487 by Duke Albrecht IV and extended to Bavaria in 1516 by Duke William IV at the Diet of Ingoldstadt, from where it spread to cover most of Germany. As well as securing revenue, the purity laws were considered a guarantee of quality.

It is probably less well known that the United Kingdom had its own purity laws. Malt was first taxed in 1660 in the reign of Charles II, but it was in the eighteenth century that legislation prohibiting the use of brewing materials other than malt and hops was introduced. These laws remained in place until the early nineteenth century when poor barley harvests and the availability of sugar imports allowed a temporary relaxation in the law.

In 1847 the law was repealed to allow up to 25% of the grist (fermentable extract) composition to be made up by sugar on a permanent basis. The principle reason was to put the sugar cane grown in the colonies on the same footing as home grown barley. The sugar was taxed in a similar manner to malt. However, the repeal of the purity law was followed by a strong lobby by the barley growers for its reinstatement, with opposition from the brewers who formed a 'Free Mash Tun Association', perhaps the forerunner of the Brewers Society.

The arguments were brought to an end in 1880, when the Free Mash Tun Act was passed repealing the duty on malt and imposing duty on the volume and specific gravity before fermentation. To quote Mr Gladstone's own words, "The brewer will brew from what he pleases and have a perfect choice of his material and his methods".

The replacement of malt with sugar and the use of sugar for priming was welcomed at the time since it enabled brewers to reduce fermentation time as well as improving the general quality of the beer. Furthermore, the large excise levies on the specific gravity of fermenting beer had a profound effect on the development of British brewing. Brewers could lower costs by reducing beer losses which carried excise duty. They also preferred to fully ferment beers to produce the maximum alcohol from the sugars present. This encouraged brewers to produce beers with lower final specific gravities and to use highly fermentable syrups to produce beers with low residual sugars to minimise duty charges.

In 1993 the excise rules changed again with excise now being raised on the percentage alcohol and the volume of beer sent out of the brewery gate. Potentially this could have as big an effect as Mr. Gladstone's 'Free Mash Tun' legislation, allowing brewers to adjust the gravity and sweetness of the beer without suffering an increase in duty unless it produced additional fermentation. As a further consequence priming sugars may find a new lease of life with the production of lower alcohol beers with more body through higher residual sugar levels.

8.3.3 Why the British favoured sugars and syrups

Whilst most other countries use cereal adjuncts such as maize, rice or wheat, the UK has chosen to use syrups made either from a starch source such as maize or wheat or as a refined syrup from cane. Why the British brewers preferred sugar is not known, but is probably due to a combination of factors, some of which are considered here.

The British generally had higher quality malts which enabled them to use single temperature mashing in mash tuns (combined malt conversion and filtration vessel) as opposed to mash kettles (multi-temperature conversion vessels) and lauter tuns (separation vessels) principally used in Europe and North America. This meant they did not need the cereal cookers available to them to use non-gelatinised starch sources.

Currently Britain still tends to use indigenous two row barley varieties, whilst Europe and America use quite a high proportion of six row barley. The six row barleys tend to produce malt with higher enzymic power thus giving them greater potential to saccharify cereal adjuncts, which in North America can be as high as 40% to 50% of the goods.

As well as the difference in enzymic power, Britain tends to produce more dark beers such as ales and stouts. Ale malts due to their higher curing temperatures have a lower enzymic power when compared with lager malts made from similar two row barley. The lower ale malt enzymic power will serve to reduce further the opportunity for solid adjuncts in the mash tun.

8.3.4 Use of syrups in brewing

Until the 1970s around 70% to 80% of beers produced in the UK were ales or stouts, and these beers have a more robust flavour and are less likely to be affected by other flavours introduced from the semi-refined cane syrups. Some of the syrup flavours are described as 'luscious' and help to give the beer a unique fullness and flavour. With darker beers the colour of the syrup was unimportant, since it could be compensated by the grist composition. Sugars and caramels are used to enhance and adjust colours. Until the 1970s major brewers were individually colouring cask beers for particular customers and this practice may still continue among some of the smaller brewing companies.

Britain is the principal producer of cask conditioned beer, which requires a secondary fermentation in the cask to produce carbon dioxide before the beer is ready for consumption. During fermentation the malt sugars are taken up in a very specific order with sugars such as glucose and sucrose taken up first, followed by maltose and then maltotriose, so that by the end of a regular fermentation only a few slowly metabolisable sugars remain to generate the secondary fermentation. Brewers in the 1840s found that adding cane priming sugars (sucrose) speeded up the secondary conditioning and made the beer ready for trade more quickly.

Syrups also have a considerable benefit in that they can be added directly to the wort kettle, and hence supply fermentable sugars without using mash tun capacity. This can increase the brewhouse throughput without the need for additional capital investment. Because syrups have a very high specific gravity, they can be used to increase the wort kettle gravity, which has a number of practical benefits to the brewer:

- More fermentable sugars can be extracted from the mashing process, thus improving yield without prejudicing the wort kettle specific gravity.
- Traditionally high gravity beers such as barley wines, which previously had to be boiled for several hours to achieve the wort kettle specific gravity, could have their boiling times reduced by making up the gravity with syrup.
- The use of syrup to increase gravity enabled the UK to develop high gravity brewing, with the benefits of increased capacity without additional capital investment or loss in brewhouse yield.

8.3.5 *Designing syrups to meet the brewer's needs*

The use of acid hydrolysis of starches to produce crude chip glucose was not readily available until the 1840s, when a typical syrup would contain up to 40% glucose, the balance being made up of higher sugars. As starch enzyme technology developed from the 1920s onwards syrup manufacturers were able to control the hydrolysis of the starch molecule and today are able to produce a complete range of different sugar spectra.

It was not until the 1960s which brought the need to brew more delicately flavoured products such as lagers, that starch-based syrups were more widely used. However, the range of products from highly fermentable glucose syrups with a similar fermentability to all malt derived beers have encouraged brewers to use syrups as a way of controlling the fermentability of their beers.

Table 8.2 shows examples of the range of carbohydrate composition of selected starch-based syrups with a breakdown into monosaccharides (DP1), disaccharides (DP2), trisaccharides (DP3) and dextrins (DP*n*).

Table 8.2 Carbohydrate spectra of syrups used for brewing

Sugar	Wort	63 DE	Maltose	VHM*	Dextrose	Malto-dextrin
Dextrose (%)	13	38	2	3	94	0.2
Maltose (%)	52	33	55	71	3.5	1.5
Maltotriose (%)	12	6	16	16	Trace	3.5
Higher sugars (%)	23	23	27	10	2.5	94.8
Fermentability (%)	71	74	65	95	98	4
Extract		270	300	310	256	

* VHM – very high maltose.

Syrups have traditionally been added directly to the wort kettle where they behave as follows:

- Contribute flavour and aroma to beer either through their own character or by diluting those flavours contributed by other grist components such as the malt.
- Improve beer stability by diluting the non-starch constituents of wort such as proteins and polyphenols which contribute to beer haze.
- Subject to fluctuations in malt prices they can often represent a cheaper source of fermentable sugars.
- Extend the brewhouse capacity by increasing throughput without capital investment.
- Add directly fermentable sugars for better brewhouse efficiency and high gravity brewing.
- Depending on the syrup used it may either enhance or dilute wort and beer colours.
- Syrups can also reduce the concentration of wort nutrients such as nitrogen, amino acids and minerals. Whilst a high adjunct level can restrict fermentation and possibly reduce the amount of foam-forming compounds, lower nutrient levels help to reduce ester formation. A lower wort nitrogen-to-carbon ratio produces lower esters, thus adjuncts can be used in high gravity brewing to avoid overproduction of esters.
- Different syrups have different fermentabilities which may be used to control the fermentability of a beer.

The way in which syrups can be used to control the fermentability of a beer is shown in Figure 8.4 where 60% of the fermentable sugar comes from malt wort and the balance is made up by the three different syrups. The effects on the fermentability as can be seen are quite marked.

The very low fermentability syrups such as the maltodextrin syrups described in Table 8.2 can be used in the production of low alcohol beers providing body and mouthfeel but very limited fermentable extract.

8.4 Summary

Sugars and syrups have played an important part in developing and retaining the character of British beers and brewing technology. The recent changes in the way excise duty is raised is likely to generate renewed interest in the use of syrups. Modern cereal adjunct production processes allow a closely controlled carbohydrate composition to be formulated, facilitating specific fermentability and sweetness/mouthfeel characteristics.

Lower fermentable syrups as described in Table 8.2 can be used to enhance the body and mouthfeel in beers where alcoholic strength continues to be squeezed due to excise duty considerations. Conversely,

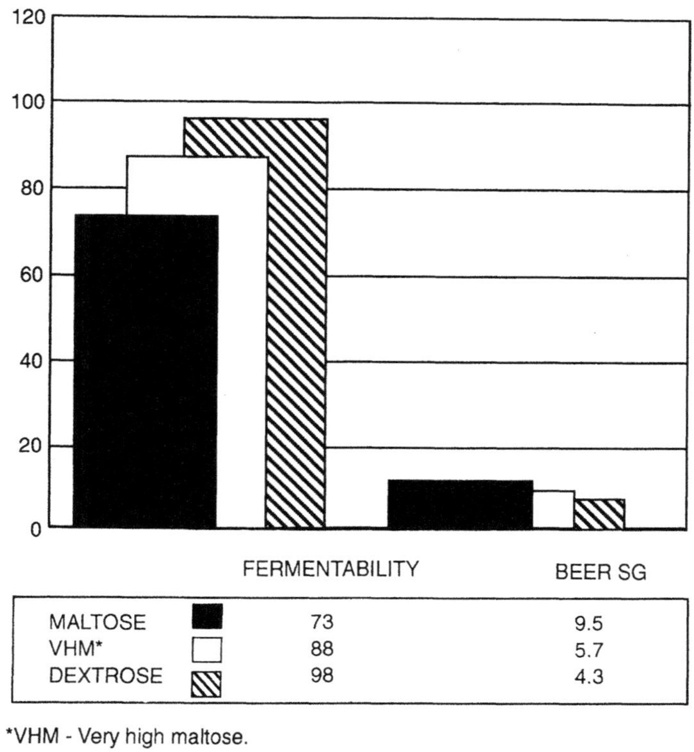

Figure 8.4 Wort fermentability using different syrups.

highly fermentable syrups with a bland flavour and very low colour have played an important part in the production of some of the light clean flavoured lagers now available.

Currently there is a new consumer trend towards more diverse products, particularly the more highly flavoured ales. This could provide a new role for darker sweeter syrups and primings to boost beer flavour and colour in a natural way. It is clear that syrups and primings have developed to such a degree that they are an indispensable raw material for the brewer, both today and for tomorrow.

References

Haarman and Reimer, *Food Ingredients Business Group Promotional Brochure.*
ICI Bio-Products Promotional Brochure, undated.
Kelco (1994) *Biospecialities Promotional Brochure.*
Mulvihill, P.J. (1992) Crystalline and liquid dextrose products: production, properties, and applications. In *Starch Hydrolysis Products*, VCH.
Rhodes, A. and Fletcher, D.L. (1966) *Principles of Industrial Microbiology*, Pergamon Press.

Bibliography

Harison, J. *Old British Beers and How to Make Them*, published through the Durden Park Beer Circle.
Jeffery, E.J. (1956) *Brewing Theory and Practice*, third edn, Adlard & Son.
Patton, J. (1989) *Additives Adulterants and Contaminants in Beer*, Patton Publications.
Wilson, J. (1992) Brewing sugars: the versatile adjuncts. In *Food Manufacturer* September.

9 Use of glucose syrups in the food industry

E.B. JACKSON

9.1 Introduction

Glucose syrups have a well established role as key ingredients in the food industry. Because of their competitive price and specific functional properties they are now used extensively in numerous applications across the whole sector. Glucose syrups are defined by the EC and Codex Alimentarius as follows:

- *EC* – Glucose syrup is a refined, concentrated aqueous solution of D-(+)-glucose, maltose and other polymers of D-glucose obtained by the controlled partial hydrolysis of edible starch.
- *Codex* – Glucose syrup is a purified concentrated aqueous solution of nutritive saccharides obtained from starch.

Glucose syrups are manufactured from starch. In Europe wheat starch and maize starch are the most common raw materials, though potato could equally well be used. The starch source will inevitably be governed by local availability and price. Whatever the source of starch, it comprises two distinct fractions, amylose, in which the anhydroglucopyranose units are linked exclusively in the α,1-4 positions, giving a linear molecule, and amylopectin, which is branched due to the presence of α,1-6 links. The molecular weight and ratio of these two fractions determine the rheological properties of the starches but have only a limited effect on the hydrolysis of starch. High amylopectin starches yield better clarity in very low DE (dextrose equivalent) maltodextrins, but otherwise the glucose properties are similar.

Starch hydrolysates of all types are produced by acid or enzymes. Figure 9.1 shows the range of products which are commonly available commercially.

9.1.1 *Product requirements*

Product types are generally classified as carbohydrate sweeteners and, as an introduction, it would be worthwhile to consider the use and selection of such sweeteners which are used to meet one or more of the requirements outlined here:

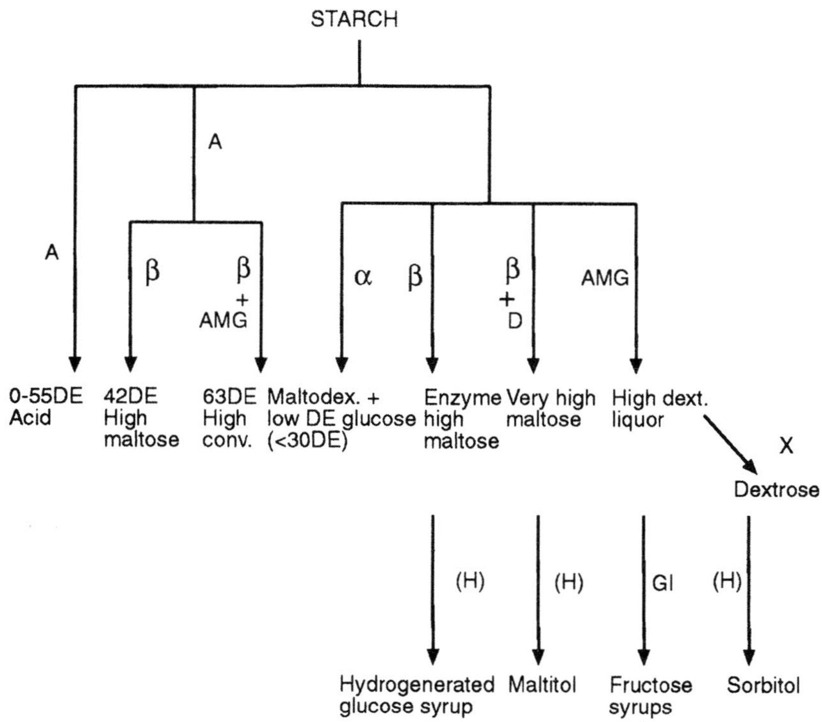

Figure 9.1 Starch products for the confectionery industry excluding blended products. A, acid conversion; α, α-amylase; β, β-amylase; AMG, amyloglucosidase; (H), hydrogenation; X, crystallistion; GI, glucose isomerase; D, debranching enzyme.

- Creation of the sensation of sweetness on the palate. Sweetness is one of the four prime tastes and foods producing this response are readily acceptable as food. It can also be linked to social behaviour where sweetness is equated to pleasurable experience.
- Product preservation, by generating a high osmotic pressure and thus inhibiting the development of microbiological spoilage.
- A source of nutritious carbohydrates. A characteristic which is particularly important in human digestion.
- Imparting a characteristic texture to the food preparation, to render it both attractive and palatable.
- A readily available source of energy.

9.1.2 Sweetener selection

Provided the sweetener has qualified as acceptable by meeting one or more of the above needs, differentiation and selection of a particular sweetener is then dependent on:

- Relative economics with the object of achieving minimum raw material costs. It is worth noting here that the price of sucrose in the EC is high and there is therefore considerable economic incentive to use glucose and/or dextrose as either an alternative or as an extender.
- Meeting existing end product requirements, or additionally meeting forward trends.
- Technical limitations, particular in respect of relative sweetness and product state.
- Legal limitations. These are currently of considerable importance for artificial sweeteners and becoming increasingly important with respect to natural sweeteners.

The physical and various functional properties of glucose syrups which influence the choice of a particular grade for a given application, are largely dependent on the dextrose equivalent (DE) of the syrup, but also in part on the carbohydrate spectra and purity.

The dextrose equivalent or reducing power of glucose syrup is commonly expressed as dextrose on a dry matter basis. Dextrose is a term synonymous with D-glucose. It was the name originally given to this saccharide and was derived from the optical properties of the saccharide. We use it today to distinguish the monosaccharide from the term 'glucose syrup' which may contain as little as 1% dextrose.

9.2 Properties of glucose syrups

Figure 9.2 gives a comprehensive list of the properties of glucose syrup as a function of the DE. Five of the more important points are discussed here.

9.2.1 Browning reaction

This is basically a condensation reaction between amino acids and reducing sugars, generally called non-enzymic browning or Maillard reaction. The reaction is enhanced by glucose syrups containing high levels of dextrose.

9.2.2 Crystallisation control

Crystallisation control of sucrose and other sugars such as lactose can be effected by means of glucose syrup. Low DE glucose syrups, because of their increased viscosity, have the greatest effect in slowing down the rate of crystallisation.

9.2.3 Humectancy

Humectancy is the ability to resist changes in moisture content, either by drying out or by picking up water. Compared to sucrose, glucose syrups are

Property or functional use	Type of syrup *Lower DE*	Type of syrup *Higher DE*
Bodying agent	←————	
Browning reaction		————→
Cohesiveness	←————	
Colour stabilisation		————→
Crystallisation control	←————	————→
Emulsion stabiliser		←————
Fermentability		————→
Flavour enhancement		————→
Flavour transfer medium		————→
Foam stabiliser		←————
Freezing point depression		————→
Humectancy	←————	————→
Hygroscopicity		————→
Increased vapour pressure		←————
Nutritive value	←————	————→
Osmotic pressure		————→
Preservation		————→
Prevention of coarse ice crystals		←————
Prevention of sucrose crystallisation		←————
Sheen producer	←————	————→
Solubilising effect	←————	————→
Sweetness		————→
Thickening Agent		←————
Viscosity		←————

Figure 9.2 Technological properties and functional uses of glucose syrups.

composed mainly of high molecular weight carbohydrates and at a given weight contribute fewer molecules. The equilibrium relative humidity (ERH) is therefore increased in food products where glucose syrup is employed. It therefore follows that high DE glucose syrups with a lower average molecular weight are more hygroscopic than low DE glucose syrups.

9.2.4 Sweetness

The sweetness of glucose syrup is generally expressed as relative sweetness, that is, as compared to sucrose, which is arbitrarily taken as 100. Many studies on the relative sweetness of glucose syrup have been conducted. The main conclusion emerging from these studies is that the relative sweetness of glucose syrup, like that of other natural sugars, is not a constant value but increases with increasing concentration. When a glucose syrup is mixed with sucrose, there is a synergistic sweetening effect, that is, their combined sweetness is greater than would be the case if they were used individually.

9.2.5 *Viscosity*

This property is basically dependent on the average molecular size. The most viscous syrups are those with the lowest DE and, conversely, the least viscous are the high DE ones. Two characteristics of glucose syrup viscosity are its sharp response to temperature and to concentration. A few degrees temperature change can drastically alter the glucose syrup from a thick and viscous material to a highly pumpable, manageable product. Slight dilution also has considerable effect on viscosity.

The viscosity of glucose syrup (at constant solids content) depends upon its chemical composition, and particularly upon the percentage of higher carbohydrates present. Glucose syrups of lower viscosity such as the 63 DE types, will facilitate the easier deposition of viscous products and can enable deposition to be carried out at higher final total soluble solids, reducing the drying time of many stoved products. If a slightly more moisture-retaining 'doctor' than 40 DE glucose syrup is wanted then 63 DE glucose syrup will be suitable.

A further point to consider is the relative sweetness of glucose syrup. It is very difficult to make a judgement on such a subjective matter. This is because the relative sweetness of different carbohydrates alters as the concentration of the solution increases and differences in sweetness become much less discernible at high concentrations. This is probably due to the fact that the high osmotic pressure of concentrated solutions dulls the sense of taste. In addition, large 'doses of sweetness' appear to have a nulling effect on this sense.

In addition, the presence of saliva results in a very rapid dilution effect. For 15% solutions, for example, 42 DE glucose syrup solids are equivalent to 36% the sweetness of sucrose and 63 DE glucose syrup solids are equivalent to 56% the sweetness of sucrose. This means that 100 parts of 63 DE glucose syrup can replace (both as regards doctoring action and sweetness) 70 parts of 42 DE glucose syrup and 30 parts of invert, a fact which is most useful where invert sugar is used.

For the food manufacturer probably the most important properties are those of viscosity and sweetness. As mentioned above the viscosity of glucose syrups decreases with increase in DE whilst sweetness increases with both DE and concentration. Even at maximum conversion, glucose syrups are only about half as sweet as sucrose at the same concentration. Synergistic effects are known to occur between the components of glucose syrups and also in glucose syrup/sucrose mixtures where the combined carbohydrates have greater sweetness than the sum of the individuals and approach the sweetness of sucrose alone.

An important trend to be seen from Table 9.1 is that the mean molecular weight decreases with increase in DE. This information can be of great value to the food formulator since he can make a theoretical estimate of

Table 9.1 Composition of glucose syrups

Type of syrup	Average molecular weight	Carbohydrate composition (%)				
		Glucose	Maltose	Maltotriose	Maltotetraose	Higher saccharides
42 DE acid converted	450	18.5	14.5	12.0	10.0	45.0
42 DE acid–enzyme converted	345	5.9	44.4	12.7	3.3	33.7
42 DE enzyme–enzyme converted	–	2.5	58.0	18.0	2.0	19.5
64 DE acid–enzyme converted	280	37.0	31.5	11.0	5.0	15.5

the blend of glucose syrups which could replace another carbohydrate of uniform molecular weight, for example, the replacement of sucrose in a frozen product such as ice cream.

9.3 Confectionery

9.3.1 High boiled sweets

To produce high boiled sweets with a satisfactory shelf life the final product must contain a minimum amount of residual moisture and the correct balance between sucrose and the glucose syrup/invert sugar components. To achieve this, sucrose and glucose syrup/invert sugar solutions are heated and boiled down to a set solids or moisture content, appropriate flavourings, colours and fruit acids being added during the boiling process. The boiling point is obviously affected by the solids level and the higher the boiling point, the higher the residual solids in the product. Under vacuum, the boiling point of an equivalent solids level or concentration is reduced and this principle is used in the food industry to some advantage. It is not possible to produce high boiled confections of adequate shelf life from sucrose solution alone and glucose syrup/invert sugar must be added for the reasons discussed below.

The main problem during the storage of high boiled sweets is the formation of small crystals, commonly known as graining which renders the product unacceptable. This is due to the recrystallisation of the sucrose, which in its simplest form may be considered as the migration of molecules to a nucleus.

High boiled sweets are a classic example of a product in the 'glassy' state. In appearance they are solid, but actually they are supercooled, non-crystalline liquids, which are so far below their melting or softening point that they have assumed solid properties without crystallising. They can be considered as liquids with very high viscosities, a property which interferes considerably with the process of crystal formation. For crystallisation to occur there must be a 'nucleus', that is, a completely sub-microscopic crystal to act as a 'seed'. These nuclei are formed spontaneously if the supersaturation is sufficiently high, but the higher the viscosity the slower the rate at which they form. Molecules of the substance crystallising have to locate and stick to the nucleus, being brought to it by their continuous, rapid vibrational movement. This movement is severely limited in a solid or in a liquid of extremely high viscosity.

To obtain a product of satisfactory texture and shelf life a solids level of 97% must be achieved, however, such a system based on sucrose alone would be supersaturated with respect to this carbohydrate making it unstable. This problem can be overcome by using, for example, a 60:40 blend of sucrose and glucose syrup. The glucose syrup increases the viscosity of the confection in its 'glassy' state, and thus reduces the mobility of the lower molecular weight carbohydrates. The viscosity of glucose syrup (at constant solids content) depends on its composition and particularly on the concentration of high molecular weight components. At the solids content of high boiled sweets, the effect of the higher saccharides is significant. Furthermore, the viscosity of the boiling mass in the pan and during its subsequent handling can be significantly affected by quite small variations in the amount of high molecular weight components in the syrup. In addition this variation can also affect the 'break' of the sweet in the mouth, that is, its brittleness and toughness.

The stability with regard to moisture pick-up or loss by any food product depends upon its ERH or relative vapour pressure. The ERH of high boiled sweets is around 30%, and since atmospheric humidity is nearly always above this there is always a tendency for the sweets to absorb moisture. Initially this only takes place on the surface; a very thin film of a solution with a solids content significantly lower than that of the remainder of the sweet is formed, which also has a much lower viscosity. In this film the reduced solids level can permit the crystallisation of sucrose (graining) to take place, which can then spread into the body of the sweet. There is some evidence to show that the presence of glucose syrup can impart a 'skin' to the outer surface of the sweet, which greatly inhibits the penetration of moisture to the interior. Glucose syrup, when used for the above reasons, is referred to as a 'doctor'. Invert sugar can also be used for the same purpose but is not so efficient.

Certain high boiled sweets depend upon graining actually being induced to produce a particular texture, e.g. rock. The mechanical action of pulling is used to beat in air and precipitate very fine sucrose crystals. Fondant is also a grained confection in which the amount of 'doctor' is much lower and the viscosity is higher because of the increased solids content.

The production of the recently available, all-enzyme processed glucose syryps presented the food manufacturers, in this case of high boiled sweets, with some interesting raw materials, in particular, high maltose syrups which are characterised by their low glucose and high maltose contents and low viscosities compared with acid and acid–enzyme syrups of the same DE. The effect of the different carbohydrate spectra of 42 DE glucose syrups on their viscosities is shown in Figure 9.3.

A useful application of these differing viscosities is that the viscosities of high boiling sugar (candy) masses of given sucrose composition can be reduced in the manufacturing process giving a better deposition performance and better (lower) air retention. Other advantages are that the mass is easier to work on a slab, does not form a hard outside crust so easily, and the final sweets are more plastic during forming with better surface flow, resulting in a higher gloss and improved finished appearance.

Conversely, similar handling characteristics can be obtained at high

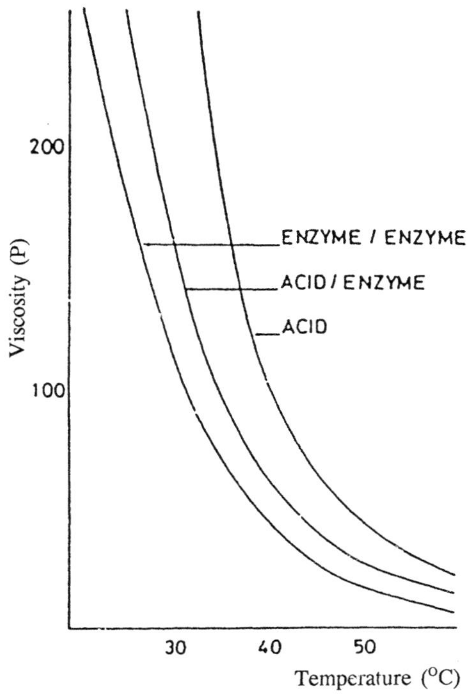

Figure 9.3 Viscosities of 42 DE glucose syrups made by different conversion methods.

glucose syrup concentrations as shown in Figure 9.4. Another useful effect is that the boiling point elevation is influenced by the glucose syrup concentration and type, illustrated by Figure 9.5. This shows the relationship between boiling temperature, at constant pressure, and residual solids for different ratios of sucrose and glucose syrups. It can be seen that by using the correct glucose syrup and syrup ratio, the boiling temperature can be reduced by as much as 5°C. This becomes of increasing importance to food manufacturers as the cost of energy rises.

High maltose syrups show less brown colour formation on heating as a direct result of their low glucose content. This is important in boiled sweets where discolorations are undesirable. The colour developed on heating glucose syrups depends on a number of factors. The first is the concentration of trace colour bodies or colour precursors, for example hydroxy methyl furfural (HMF) and amino acids which may be present in the syrup, and the second is the reducing sugar content. The majority of colours are produced as a result of the Maillard reaction and simple sugars, such as glucose, undergo the reaction most readily.

9.3.2 Production of high boiled sweets

There are three main production methods used for high boiled sweets, namely open pan, vacuum cookers and continuous cookers. Each of these requires a different optimum ratio of sucrose to glucose syrup to give the best results and to maintain product quality by preventing sucrose crystallisation. The reason for an increased glucose syrup content in boiling

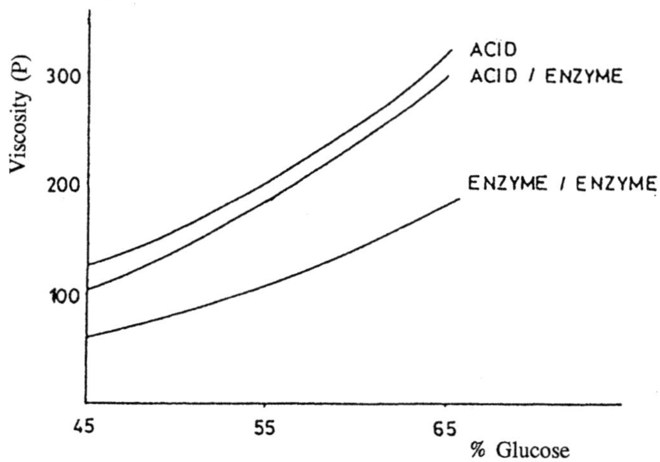

Figure 9.4 Candy viscosities of various mixtures of sucrose and glucose syrups.

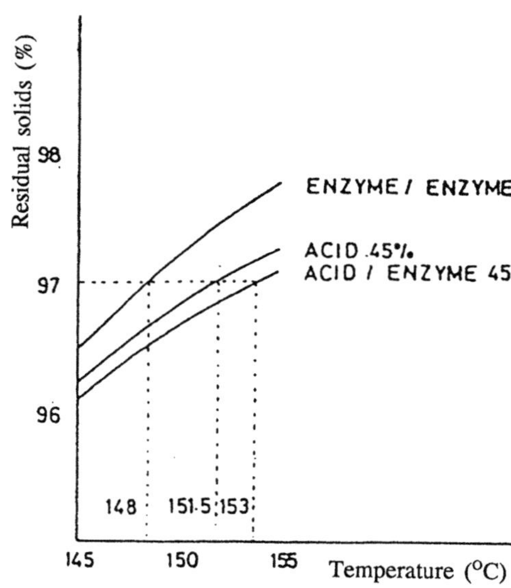

Figure 9.5 Effect of glucose syrups in candy masses on boiling point (Howling, 1979).

is twofold. First, as methods become more advanced the amount of agitation to which the syrup is subjected is greatly increased, and second, the amount of time for which the syrup is subjected to high temperatures is considerably reduced, thus lessening the danger of inversion. It should be noted that these comments only apply to the more traditional glucose syrups, such as the acid and acid–enzyme converted products rather than the new enzyme–enzyme syrups.

Open pan boiling is limited in application since at the high temperatures involved, discoloration of the syrup may occur. Vacuum cooking overcomes this problem since the temperature at which the same total solids content is obtained is lower. Open pan methods can produce syrups containing 95–96% solids whereas with vacuum methods a total solids of around 98–99% can be achieved. The higher the solids content the longer the shelf life of the product. With vacuum cooking excessive agitation of the syrup at the predetermined temperature for sweet production should be avoided since this may induce sucrose crystallisation. Syrup feed rate, steam pressure, and precook temperature should remain constant to maintain the quality of the product.

Continuous methods of manufacture often use a microfilm-type cooker which is basically a scraped film evaporator. Heat is supplied by superheated steam and sucrose/glucose syrup solution is introduced continuously at the top of the cooker. The sugar is cooked in a turbulent film about 1 mm thick in about 8 s and is not discoloured by heating.

After cooking the sugar masses from the above processes are discharged and further processed into their final forms ready for distribution and sale.

9.3.3 Toffees, caramel and fudge

This family of confectionery products is characterised by the development of colour and flavour by the encouragement of the Maillard browning reaction with the milk proteins. As a rule therefore acid-converted syrups containing higher levels of dextrose are preferred but it is possible to alter processing conditions to accommodate high maltose syrups in some cases particularly where their lower relative candy viscosity is desirable in high speed processes. Also a lighter colour can be desirable in some instances to give a creamy appearance. The difference between caramels and toffee lies normally in the moisture content of the products and their fat content. Both are essentially complexes of fat globules surrounded by a concentrated sugar solution containing the non-milk fat solids. The sucrose/lactose/glucose syrup ratio is of prime importance with regard to the keeping properties of the product. Acid-converted 42 DE glucose syrup is usually used although viscosity, colour development and firmness of the product may be changed by using a different glucose syrup. The higher the concentration of the glucose syrup the greater the stretch properties of the

Table 9.2 Typical process for making slab toffee

Proportion required (no. of parts)	Ingredient
4	Brown sugar
10	White granulated sugar
14	Glucose syrup
10	Sweetened condensed milk
10	Hard toffee fat (in small pieces)
2	Dairy butter
0.015	Glyceryl monostearate or lecithin

Stir as rapidly as possible for 5–10 min using just sufficient heat to melt the fat to premix before raising the heat for actually cooking. Much care and good judgement as to the intensity of heat required and continuous stirring throughout the cooking operation will be required.

Cook until a sample extracted by means of a wetting rod which is then plunged into cool water and worked by the fingers proves to be sufficiently firm and hard. Then turn off the heat, add a sufficient quantity of salt and flavour as preferred, stir in briefly, at the same time adding nut or raisin fillers if required. Pour out immediately. Temperature should be approx. 250°F (121°C).

caramel and toffee. Higher DE syrups increase flowability and maltose syrups produce caramels which are light in colour and creamy in appearance.

The presence of milk solids causes the product to be different in its properties from other types of confection. During processing the milk protein and the reducing sugars may undergo the Maillard reaction, the resulting compounds contributing to the overall colour and flavour of the caramel or toffee. The reaction is fastest at temperatures in excess of 95°C, and the higher the temperature and the longer the time of heating the greater the flavour development and darker the colour. Caramelisation of the sugars also occurs at high temperatures and the end products of this reaction also contribute to the caramel flavour and colour. A typical caramel formulation is shown in Table 9.2. The fat is usually a mixture of hardened palm kernel oil and butterfat. Small amounts of salt and flavouring are also usually added.

9.3.4 Gums and jellies

This is a popular class of confection in the USA and Europe and is gaining popularity in other parts of the world. Basically these are low boiled confections with a residual moisture of about 20% and whose major textural characteristics are given by the hydrocolloid used. Hydrocolloids are commonly starch, gelatine, agar, pectin and gum arabic. With the exception of gum arabic in gum confections where it is used at 35–45%, the hydrocolloid comprises less than 10% and the balance is largely glucose syrup and sucrose. Although the major properties are exerted by the hydrocolloid, selection and level of glucose syrup still has a part to play in the properties of this class of product since it is included at 30–50% of the formulation.

Traditionally 42 DE acid-converted glucose syrup is used in these products to give protection against crystallisation of the sucrose. If, however, tender eating properties are required the gel properties can be improved by the use of enzyme high maltose syrups where the reduction in higher sugar causes less stickiness and a cleaner gel performance. The higher sugars cause problems with pectin and gelatine gels as they interfere with the macromolecular set of the hydrocolloid.. Where starch is the hydrocolloid, this of course does not apply and in starch-based pastilles, for example, 36 DE glucose syrup is used to assist the viscosity of the system. In pectin jellies where 42 DE glucose can cause toughness in the gel, again high maltose syrup or even 63 DE glucose syrup may be used to overcome this and also to give increased sweetness in the case of the 63 DE glucose syrup.

9.3.5 Marshmallow

Marshmallow is an aerated sugar-containing foam, stabilised with gelatine and egg albumen. Gum arabic and starch may be added. It is produced by a cold process in which a sugar syrup (previously prepared by boiling) and a solution of a whipping agent are mixed and aerated. A 63 DE glucose syrup is normally used because it has good humectant and whipping properties. Invert sugar, glucose or sorbitol may also be added. High maltose syrups are also used owing to their low viscosity and high moisture retention properties. They will also help prevent glucose crystallisation.

9.3.6 Nougat

Nougat is basically a high boiled syrup containing fat to which is added a frappe, containing either egg albumen or gelatine. Whipping agents, starch, gum arabic or agar may also be added as texture modifiers. For chewy, non-grained, non-chocolate enrobed nougat, a 36–38 DE or high maltose syrup is used to maintain viscosity in the syrup phase and help prevent stickiness. For non-enrobed, short textured and grained nougat a traditional 42 DE glucose syrup is used to increase moisture retention. For chocolate covered nougat a 63 DE syrup is used to provide considerably more moisture retention and reduce drying out on storage.

9.3.7 Chewing gum

Chewing gum is another confectionery item widely produced both in Europe and the United States. The most common form is the stick pack.

Increasingly sugar-free chewing gum is being produced using sugar alcohols. Traditional chewing gum consisted of a sugar impregnated into gum base. Glucose syrup typically provided some 25% of this sugar. The type of glucose syrup used was acid-converted 42 DE or 35 DE, the higher sugars adding to the chewiness of the texture and the slow and sustained release of the sugar. The most significant factor which characterises glucose syrup for chewing gum is the concentration, usually 44–45°Bé and 85% dry solids. This is because the chewing gum process is not a boiling process and any water added has to be minimised.

9.4 Dextrose monohydrate in sugar confectionery

Dextrose is a natural sweetener much valued as an ingredient in confectionery applications. It can be easily handled, weighed, mixed and dissolved. However, dextrose is much more than an 'alternative' sweetener, due to its functional properties.

9.4.1 Sweetness

In an arbitrary scale of sweetness where sucrose is given a value of 100, dextrose is rated at 75. It can therefore be used to reduce sweetness. This, however, is an oversimplification because flavour, acidity and presence of other sugars at total concentration influence the relative sweetness. For example, when used with sugar at up to 40% replacement level in a formulation of high sweetener concentration and low acidity the relative sweetness of dextrose could be as high as 90 as a result of the synergistic effect which occurs when these two sweeteners are blended. This technique of using dextrose in combination with sugar is useful in achieving flexibility of sweetness balance.

9.4.2 Crystallisation and solubility

Dextrose is less soluble than sucrose at room temperature and hence is more prone to crystallisation (Table 9.3). However, in a combined sweetener system there is an increase in overall solubility. The total solubility in a given amount of water is greater than the sum of the individual components. Crystallisation is therefore inhibited or delayed. As a result, in highly oversaturated systems (e.g. fondant) dextrose has a marked softening effect upon the texture.

The heat required to dissolve dextrose (25.2 cal g^{-1}) is much greater than that for sucrose (3.85 cal g^{-1}). In consequence when eating an item of food containing dextrose in the crystalline state, there is a distinct cooling sensation in the mouth. This significantly increases the product appeal of

Table 9.3 Solubility of dextrose and sucrose (g per 100 g water)

Temperature (°C)	Dextrose*	Sucrose
10	65.2	190.5
30	120.5	219.5
50	243.5	260.4
70	359.3	320.5

* Calculated as anhydrous

fillings and icings, for example, and particularly those flavoured with peppermint or lemon.

9.4.3 Molecular weight and osmotic pressure

Dextrose is a monosaccharide and has a molecular weight of 180 compared with that of 342 for sucrose which is a disaccharide. This difference affects the properties of the sweetener in solution. For example, at any given concentration, a dextrose solution contains almost twice as many dissolved molecules as a sucrose solution. As a result dextrose exerts a greater osmotic pressure, an important consideration where preserving action is being sought.

9.5 Dextrose application in sugar confectionery

9.5.1 Toffees, caramel and fudge

In this type of product, the use of dextrose will yield a colour-forming effect, coupled with an increase in flavour. A side effect of using dextrose in compound formulations is that the viscosity of the cooked toffee in the dipping pots is stabilised, thereby reducing waste caused by the material becoming too heavy to work. An upper limit of 10%, based on weight of sucrose should be used.

9.5.2 Powdered sherbet centres and lemonade powder

Lemonade powders and powdered centre fillings for hard boilings may be formulated with dextrose as the sole sweetener. Because dextrose dissolves instantly it gives the impression of a liquid centre and has a pleasantly cooling effect on the tongue. Sucrose may be introduced if an increase in sweetness is desired. It is also possible to produce a powdered filling, based on dextrose, with reduced levels of acid and flavour.

9.5.3 Chewing gum and bubble gum

In sugar coated chewing gum and bubble gum, up to 10% of the icing sugar normally used can be replaced by dextrose monohydrate.

9.5.4 Tablets

Unlike sucrose, dextrose may be used for direct compression of tablets merely by the addition of lubricant (1% magnesium stearate) but without recourse to granulation or other pretreatment. The resulting tablet has cool eating properties and good texture.

If desired, dextrose may be used with sugar at any level of combination. In this case, dextrose is blended with the standard granulated mix.

9.5.5 Dextrose fondant

Dextrose fondant is a novel concept in which the product is formulated to contain not more than 20% sucrose. The resulting fondant has an ultra-smooth texture and a pronounced mouth-cooling effect upon eating. In spite of containing about 7% less solids than traditional fondant, the dextrose-based product is completely stable microbiologically. Dextrose fondant has adequate soluble solids in the syrup phase, but yields the desired texture with a much lower crystal phase than is found in normal fondant. Apart from cost saving on ingredients, dextrose fondants can be prepared without boiling and without specialised fondant machines.

9.6 Preserves

Preserves are a food commodity which require and use large quantities of sugar and glucose syrup. During the manufacturing process fruit is boiled with the sugars to a solids content which will give stability to microbiological spoilage, and to the state where provided other factors such as pH are correct the sugar–acid–pectin systems forms a stable gel. Glucose syrups can and are used to replace up to 50% of the carbohydrate content. The most suitable and commonly used is a 63 DE glucose syrup which has an average molecular weight of approximately 300 Da, which provides a higher osmotic pressure than the equivalent weight of sucrose, and thus will contribute an improved effect on the microbiological stability of the preserve.

The presence of glucose syrup in preserves will also inhibit the recrystallisation of the lower molecular weight carbohydrates such as the sucrose. Sucrose is saturated at 67% solids, so if it were used alone in the commercial preserves, crystallisation would occur, thus detracting from the organoleptic acceptability of the finished product. Such crystallisations can be prevented by the addition of a second sugar. Traditionally naturally inverted sugar was used which readily occurred in the more old-fashioned type of open pan boiling. However, with more modern vacuum techniques being used this natural inversion is minimised. Even if natural inversion or added invert sugar is used the maximum solids which can be attained before crystallisation takes place is 76%. This produces only a fairly limited safety margin in commercial preserves, and is not suitable where high solids are used, for example in bakery jams. The inclusion of glucose syrup reduces this risk of crystallisation.

When more than 50% of the carbohydrate portions in a preserve is glucose syrup, it is claimed that the gel strength is weakened and an undesirable texture (mouthfeel) is produced. This is thought to be due to the higher molecular weight sugars present and appears to be the limiting factor in the level of glucose syrup which can be incorporated. However, the new higher dextrose and higher maltose-containing syrups which automatically have a reduced proportion of higher sugars, should theoretically be capable of being used at levels greater than 50%.

It has been claimed that the inclusion of glucose syrup increases damage caused by syneresis, however, the evidence is not conclusive since studies have demonstrated that whilst glucose syrup affects sugar–acid–pectin systems, if alterations to the set and manufacturing conditions are made, then extra cut gel syneresis can be avoided.

Fundamental work done by D. Howling (Cerestar) in conjunction with G. Stainsby, University of Leeds, showed that the type and level of glucose affected the setting temperature of the sugar–acid–pectin system. This varied according to the pectin type and pH but could be as much as 15°C (Figure 9.6). The raising of the setting temperature was causing premature setting in the process before filling of the finished product containers and this shearing of the infant gel in fact led to the syneresis. By increasing filling temperatures to avoid premature set glucose levels up to 50% were obtained in commercial preserves and recently this has been moved higher.

The type of glucose syrup used traditionally was 63 DE acid–enzyme glucose syrup but as discussed the higher sugars do tend to give a sticky, viscous non-elastic gel. Higher 68 DE enzyme–enzyme glucose syrups are now used to give superior gel properties. pH values are important in the sugar–acid–pectin system and adequate buffering must be provided in the product.

The use of all sucrose and invert sugar in preserves produces a very sweet product. This can be tempered by the inclusion of glucose syrup which allows the flavour of the fruit present to be sensed more readily by the consumer's palate.

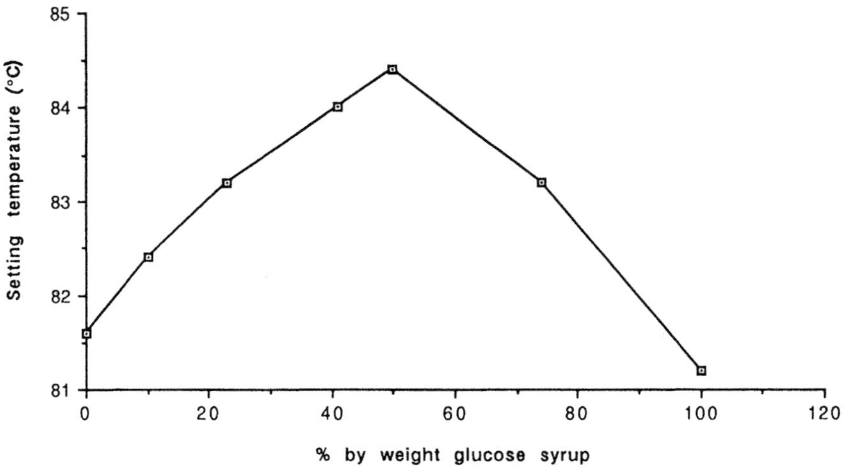

Figure 9.6 Set temperature versus glucose syrup ion sugar–acid–pectin systems.

9.7 Ice cream and frozen desserts

9.7.1 Ice cream

In general terms, ice cream may be defined as an emulsion of fat, sugar and milk solids non-fat which has been simultaneously whipped and frozen. Unless the product is described as 'dairy ice cream' the fat need not necessarily be of dairy origin, i.e. butter or cream. The legal standards of composition are contained in the 'Ice Cream Regulations' No. 1866, 1967. The consumer appeal of ice cream is greatly enhanced by the addition of flavours. Traditionally, the most important of these are vanilla, fruit, fruit extracts, coffee and chocolate.

Ice cream is presented to the consumer in a number of different ways. It is served, from mobile vans, by scoop into biscuit cones or wafers, out of small bulk containers. In recent years soft-serve ice cream has become very popular, where the product is drawn straight from the freezer for direct consumption. For domestic use as a dessert, ice cream is sold by the retailer in cartons or plastic tubs and stored by the consumer in the freezer until required. Basically the same method is used by hotels, restaurants and other institutions which serve ice cream as a dessert or part of a dessert course.

9.7.2 Milk ice

The manufacture of this product, also known as 'ice milk', is covered by the ice cream regulations, being a type of ice cream low in fat and milk solids. Unlike standard ice cream, the fat used in the manufacture of milk ice must be derived wholly from milk. Although by no means as popular as ice cream, milk ice is very refreshing and should appeal to the calorie conscious consumer.

9.7.3 *Mousse*

Mousse is a dairy food based on sugar and milk solids with a high fat content, which may be cream and/or vegetable fat. Although of high solids content, mousse is quite light in texture, by virtue of the great extent to which it has been aerated. As with most dairy products, mousse is amenable to flavouring, fruit flavours and chocolate being the most popular.

9.7.4 *Water ice and ice lollies*

Water ice and ice lollies are frozen confections that do not contain milk solids or fat. They are, in effect, frozen sugar syrups of about 15–25% solids content, usually coloured, fruit flavoured and containing a small quantity of citric acid.

Water ice and ice lollies differ in that the former is drawn from the freezer with about 20% overrun, while the latter are frozen in moulds without aeration. Ice lollies containing fruit juice are particularly refreshing.

9.7.5 *Function of glucoses in ice cream and frozen desserts*

Sweeteners are essential constituents in ice cream and frozen desserts and greatly enhance their general palatability. Their sweet taste balances the flavour of the fat and accentuates the creaminess of milk solids.

By contributing to the total solids content of an ice cream mix, sugars exert a beneficial effect on the body, texture and melt down of the product. As true solutes they also influence the freezing characteristics of the mix.

Ice creams deficient in sugars are generally of crumbly, coarse or flaky texture and lacking in sweetness. Conversely, excess sugars yield a heavy, soggy texture, as well as over sweetness, and in extreme cases may create difficulties in freezing. It is important, therefore, to strike a correct balance of ingredients in the formulation of frozen dairy products and also, in the case of ice creams, to select a total mix appropriate to the type of freezer.

Although cane or beet sugar (sucrose) is the prime sweetener used in the preparation of ice cream and frozen desserts, a valuable range of complementary carbohydrates exists in the form of glucose syrups. In combination with sucrose the glucose syrup provides significant benefits to the manufacturer by adding another dimension of flexibility in his formulations. Depending on the particular glucose syrup selected, and by utilising its specific functional properties, a much wider degree of control may be exerted on sweetness, body, melt down and texture than by the use of a single sweetener.

In order to gain the maximum benefit from the versatility of glucose syrups it is necessary to consider in more detail how they influence the properties of frozen dairy products. Their main functions are as follows:

- to control body and texture,
- to confer sweetness,
- to improve shelf life,
- to provide a nutritious source of milk solids.

9.7.6 *Selection and use of glucoses in ice cream and frozen desserts*

The carbohydrate composition and functional properties of glucoses differ markedly and largely depend upon their degree of conversion (Table 9.1). Syrups that can be used vary widely from a 20 DE maltodextrin to dextrose itself at 100 DE. Although they can be included in certain formulations to achieve specific effects, their random and haphazard use can create problems for the manufacturer. For example, a manufacturer might select dextrose monohydrate in order to retain maximum sweetness in hardened ice cream bricks. Unfortunately the effect upon the mix freezing point would probably cause unacceptable softening of the product. On the other hand its use is ideal in soft-serve ice cream and ice lollies. In ice lollies, which usually are low in solids content, dextrose monohydrate exerts a beneficial softening of texture. It also helps to prevent the surface crusting of sugar crystals and 'white spots' caused by the expulsion of dye (Table 9.4).

By virtue of its high molecular weight constituents, maltodextrin at 5% of the mix weight will give complete protection against the development of

Table 9.4 Glucose syrup type and typical inclusion rates in ice cream and frozen desserts

	Type of glucose			Recommended usage level (% total sweetener)	
Product	40 DE	63 DE	100 DE	Range	Optimum
Ice cream					
Soft serve		*	*	10–25	20
Hardened brick	*	*		10–25	20
Packaged, small bulk	*	*		10–25	20
Milk ice		*	*	10–15	12
Water ice		*	*	10–25	20
Ice lollies			*	10–25	20
Sherbet		*	*	10–25	20
Mousse	*	*		10–25	20
Frozen desserts	*	*		15–30	25

coarseness in stored ice cream, but the texture will be too 'chewy'. However, it will replace 10% of the non-fat milk solids in ice cream with improvement of texture and providing a useful cost benefit.

Generally, the most useful glucose syrups for ice cream and sherbet formulations are the 63 DE types. Their balance of carbohydrates is such that they provide adequate sweetness and yet contain sufficient 'higher sugars' (sugars of high molecular weight) to ensure smoothness of texture and protection against coarseness. Furthermore their freezing point depression is fairly close to that of sucrose giving a very slight softening effect.

When it is desirable to build up total solids and yet necessary to avoid oversweetness, 40 DE glucose syrups and high maltose glucose syrups are recommended. Their use is often preferred in sherbets, mousse and a wide variety of frozen desserts, for smoothness and bodying of texture.

9.8 Bakery products

9.8.1 Cakes and biscuits

Tests have shown that partial replacement of sugar by glucose syrup yields several advantages in baked goods. The keeping quality is improved and there is a slight increase in cake volume. The use of high DE glucose syrup is particularly advantageous in high ratio formulations where the high level of sugar frequently produces an oversweet product. Significant cost savings will also result from the use of high DE glucose syrup. Recommended levels of use are as follows (percentages represent sugar replacement):

- Madeira slab cake 10%
- High ratio slab cake 10%
- Sponge sandwich cake 10%
- Swiss roll 25%

9.8.2 Yeast-raised products

In the past, yeast-raised goods have been made primarily with sucrose or dextrose as the sugar components. The recent development of high fermentable glucose syrups of the 63 DE type has made possible the introduction of syrups into the field of yeast-raised goods. The syrups used here are designed to contain the highest possible percentage of mono- and disaccharides. These supply fermentable carbohydrates for the yeast, add to the flavour of the finished product through residual sweetness and also contribute to colour development in the crust. Regular and lower conversion syrups are not satisfactory.

9.9 Glucose syrup in savoury sauces, tomato sauces and salad dressings

Glucose syrup can be successfully employed as a partial replacement for sucrose and in many cases as a 100% replacement for sucrose, with the additional benefits of enhancing flavour, improving texture, extending shelf life and imparting a desirable gloss. The type of glucose syrup recommended is normally 63 DE glucose syrup, however, in examples where body is required low DE glucose syrup can provide additional benefits and can also reduce the percentage of stabiliser used.

9.10 Coffee whiteners

Coffee whiteners are not imitation cream. They are formulated systems similar to the natural dairy product, with the advantages of lower cost, longer shelf life, and excellent mouthfeel and physical properties. Properly manufactured coffee whiteners combine five basic components; vegetable or animal fat, protein, carbohydrate, stabiliser–emulsifier and water in a stable product with a delicate flavour that disperses quickly in coffee.

Coffee whiteners are customarily marketed either as a dry, free-flowing powder or as a liquid constituted for immediate use. A typical formulation is as follows (percentages are dry weight):

Low DE glucose syrup or maltodextrin	50–60%
Emulsifier–stabiliser	3–5%
Protein (caseinate)	4–8%
Fat (shortening or oil)	30–40%
Buffering agent	1–5%
Flavouring and colour as required	

Carbohydrates, acting as bulking agents, must be present to adjust the density and viscosity of the system by adding soluble solids to it without undesirable sweetness. The most acceptable are low DE glucose syrups or maltodextrins.

A truly bland, dairy-like flavour is important, one typically augmented by low sweetness carbohydrates in liquid or dry solids form.

Among the other desirable characteristics of a good coffee whitener are low hygroscopicity (in the powered product) to prevent lumping and caking, uniform suspension in coffees of varying pH values and quick, homogeneous dispersion in hot liquids.

9.11 Dried glucose syrups or maltodextrins (dried starch hydrolysates)

Glucose syrups can be spray dried to produce a soluble white amorphous powder which can be a more convenient form to handle than a thick syrup.

In European convention up to 20 DE glucose syrups are referred to as maltodextrins and above 20 DE as glucose syrup solids or dried glucose syrup, which is obviously only an arbitrary distinction.

The carbohydrate spectrum of these products is obviously identical to the parent syrup and, like the parent syrups, the functional properties are dependent on the dextrose equivalent. For example, lower DE maltodextrins are less sweet, less hygroscopic and less likely to form colour in the browning reaction (Figure 9.2).

A 20 DE maltodextrin is a food ingredient with an unusual combination of properties; it is a fully soluble carbohydrate of low bulk density and has virtually no sweetness. Spray dried glucose syrups offer the functional properties of liquid glucose syrup coupled with the convenience of being in a low-moisture powder form.

Maltodextrins and spray dried glucose syrups find applications in a wide range of processes and convenience foods, principally as fillers/carriers and sweetness reducer (Table 9.5). To some extent the products are interchangeable and ultimate choice depends upon the emphasis desired on specific functional properties. They are ideally suited for blending into powdered formulations, and serve the following purposes. As fillers/carriers, they can (1) reduce loss of volume in contents on storage or in transit; (2) absorb fats and oils to retain free-flowing properties; (3) aid dispersion during makeup; (4) standardise the quality of products prepared from variable natural sources (spice mixes, flavours); (5) dilute concentrated or expensive ingredients (flavours, flavour enhancers); (6) save costs by partial replacement of lactose and milk powder in certain applications, such as in the manufacture of tablets.

Many products can be made more acceptable by lowering their level of sweetness. In most cases it is essential to achieve this without reducing the total solids content. Reformulating, using low conversion hydrolysates, achieves this reduction without greatly affecting a number of important properties. These include increased chemical reducing power which produces faster colour and flavour compounds in the Maillard browning reaction when heated in the presence of an amino compound.

Table 9.5 Uses of spray dried products

1. Sweetness controller
2. Viscosity controller
3. Diluent – filler
4. Carrier for oils/fats
5. Crystallisation inhibitor
6. Convenient handling/mixing
7. Partial replacer for skimmed milk powder, lactose and whey powder

9.12 Conclusions

This chapter has set out to review objectively the role of glucose syrups in the food industry. Glucose syrups have been shown to be a highly versatile group of ingredients which have proved useful in various applications throughout a wide variety of food products.

The knowledge and experience of the food formulator can be used to take advantage of the properties of glucose syrups, not only to improve product quality but also to bring about significant savings in overall raw material and energy costs. Advances in technology will allow for further diversification as needs arise to meet the specific requirements of the customer base. An example of this is the isomerised, highly converted glucose syrup as a cost effective substitute for the traditional invert sugar used in the soft drinks industry.

Index

9 780751 402698

GPSR Compliance

The European Union's (EU) General Product Safety Regulation (GPSR) is a set of rules that requires consumer products to be safe and our obligations to ensure this.

If you have any concerns about our products, you can contact us on ProductSafety@springernature.com

In case Publisher is established outside the EU, the EU authorized representative is:

Springer Nature Customer Service Center GmbH
Europaplatz 3
69115 Heidelberg, Germany

Batch number: 10620086

Printed by Printforce, the Netherlands